ACKNOWLEDGEMENTS

We are grateful to the Science and Engineering Research Council who provided the computing facilities used for the work described here. We also wish to thank Dr. Malcolm Smith who kindly read through the manuscript and made some helpful comments.

Y.S.H and A.G.J.M

CONTENTS

NOTATION iii

CHAPTER 0 INTRODUCTION 1

CHAPTER 1 SINGULAR-VALUE, CHARACTERISTIC-VALUE AND POLAR DECOMPOSITIONS 5

1.1 System Description and Feedback Configurations 5
1.2 Characteristic Gain Loci and the Generalized Nyquist Stability Criterion 7
1.3 The Singular Value Decomposition (SVD) 10
1.4 SVD of a Continuous Matrix-Valued Function 13
1.5 Polar Decomposition (PD) 15
1.6 Normality and Spectral Sensitivity 17

CHAPTER 2 USE OF PARAMETER GROUP DECOMPOSITION TO GENERATE NYQUIST-TYPE LOCI 21

2.1 Some Matrix Groups and their Parametrizations 22
2.2 Dimension of Matrix Groups 26
2.3 Nyquist-Type Loci — the PG Loci 27
2.4 Relationship between the Parameter Group Decomposition and Normality 32
2.5 Parametrization of Higher Order Matrix Groups 34
2.6 A Drawback of the Parameter Group Decomposition 35

CHAPTER 3 ALIGNMENT, NORMALITY AND QUASI-NYQUIST LOCI 37

3.1 Frame Alignment and Normality 37
3.2 Relationship between Skewness and Misalignment 42
3.3 The Quasi-Nyquist Decomposition (QND) 42
3.4 Eigenvalue Bounds and the QND 44
3.5 Quasi-Nyquist Loci (QNL) 49
3.6 Standardization at $s=0$ or ∞ 53
3.6.1 Standardization at $s=0$ 53
3.6.2 Standardization at $s=\infty$ 54
3.7 Diagonalizing at a Critical Frequency 57

CHAPTER 4 A QUASI-CLASSICAL DESIGN TECHNIQUE 66

4.1 Computer-Aided Control System Design 66
4.2 Stability 68

4.3 Performance 69
4.3.1 Reversed-Frame-Normalizing (RFN) Controller 70
4.3.2 Interaction 72
4.3.3 Tracking Accuracy and Disturbance Rejection 74
4.4 Robustness 75
4.5 Robustness and Reversed-Frame-Normalization (RFN) 80
4.6 Compatibility Conditions 82
4.7 Specifying a Desired Compensated System 85

CHAPTER 5 CALCULATING A COMPENSATOR NUMERATOR MATRIX BY LINEAR LEAST-SQUARES FITTING 90

5.1 Reversed-Frame-Normalizing Design Procedure (RFNDP) 90
5.2 Some Results for the Linear Least-Squares Problem 93
5.3 Calculation of the Precompensator Numerator Matrix 96
5.4 Example 98

CHAPTER 6 CALCULATING A COMPENSATOR BY NONLINEAR LEAST-SQUARES FITTING 105

6.1 Problem Formulation 106
6.2 A Least-Squares Problem whose Variables Separate 109
6.3 Example 112

CHAPTER 7 EXAMPLES OF THE DESIGN TECHNIQUES 117

7.1 A Design Example for a Turbo-Generator 117
7.2 Non-Square Systems 124
7.2.1 Systems with More Inputs than Outputs 124
7.2.2 Systems with More Outputs than Inputs 126
7.3 Design Examples for Systems with More Outputs than Inputs 131
7.4 General Conclusion 148

APPENDIX A Analytic Properties of the Singular Values of a Rational Matrix 150
APPENDIX B Proofs of Prop 3.2.1, Prop 3.3.1, Prop 4.5.1 and Theorem 4.6.2 156
APPENDIX C The System AUTM 163
APPENDIX D The Systems NSRE and REAC 165
APPENDIX E The System TGEN 167
APPENDIX F The System AIRC 169

REFERENCES 171
BIBLIOGRAPHY 176
INDEX 180

NOTATION

A list of recurrent symbols is given below.

$a := b$ means a is defined to be b or a denotes b

$\mathbb{R}, \mathbb{C}$:= field of real and complex numbers, respectively

$\mathbb{C}_+$:= $\{z \in \mathbb{C} \mid \text{Re}\,z \geq 0\}$, the closed right half plane (closed RHP)

$\mathbb{C}_-$:= $\{z \in \mathbb{C} \mid \text{Re}\,z \leq 0\}$, the closed left half plane (closed LHP)

$\mathbb{C}_+^*$:= $\mathbb{C}_+ - \{0\}$

$D(c;r)$:= $\{z \in \mathbb{C} \mid |z-c| \leq r\}$, the closed disc centre c, radius r

For any $\Omega \subset \mathbb{C}$,

Ω° := interior of Ω, e.g. $\mathbb{C}_+^\circ$ denotes the open PHP

For $z \in \mathbb{C}$

$|z|$:= modulus (or magnitude) of z

$\angle z, \arg z$:= argument of z

$\text{Re}\,z, \text{Im}\,z$:= real, imaginary part of z, respectively

$\bar{z}$:= complex conjugate of z

$\mathbb{R}[s]$:= ring of polynomials in s with coefficients in $\mathbb{R}$

$\mathbb{R}(s), \mathbb{C}(s)$:= field of rational functions in s with coefficients in $\mathbb{R}, \mathbb{C}$

$\mathbb{R}_p(s)$:= $\{g(s) \in \mathbb{R}(s) \mid \lim_{s\to\infty} |g(s)| < \infty\}$, set of proper rational functions

$\mathbb{R}_{sp}(s)$:= $\{g(s) \in \mathbb{R}(s) \mid \lim_{s\to\infty} |g(s)| = 0\}$, set of strictly proper rational functions

Let $\mathbb{F}$ be any one of $\mathbb{R}$, $\mathbb{C}$, $\mathbb{R}[s]$, $\mathbb{R}(s)$, $\mathbb{C}(s)$, $\mathbb{R}_p(s)$ or $\mathbb{R}_{sp}(s)$, then:

$\mathbb{F}^{m\times \ell}$:= set of $m\times\ell$ matrices with elements in $\mathbb{F}$

$\mathbb{F}^n$:= vector space of $n\times 1$ column vectors with elements in $\mathbb{F}$, over an appropriate field

Let $M \in \mathbb{F}^{m\times\ell}$ where $\mathbb{F} = \mathbb{R}$ or $\mathbb{C}$, then:

m_{ij} := (i,j)th entry of M; we also write $M = (m_{ij})$

$\lambda(M)$:= spectrum (set of eigenvalues) of M

$\sigma(M)$:= set of singular values of M

$\sigma_{max}(M)$:= $\max \sigma(M)$, maximum singular value of M

$\sigma_{min}(M)$:= $\min \sigma(M)$, minimum singular value of M

M^T := transpose of M

M^*	:=	conjugate transpose of M
$M^\dagger$	:=	Moore-Penrose inverse of M
$\vert M\vert$	:=	(x_{ij}) where $x_{ij} = \vert m_{ij}\vert$
arg M	:=	(x_{ij}) where $x_{ij} = \arg m_{ij}$
$\mathcal{R}(M)$	:=	$\{Mx \mid x \in \mathbb{F}^\ell\}$, range space of columns of M
$\mathcal{N}(M)$	:=	$\{x \in \mathbb{F}^\ell \mid Mx = 0\}$, right null space of M
Tr(M)	:=	$\sum_{i=1}^{m} m_{ii}$, trace of M, if M is square
$\Vert M\Vert$	:=	$[\mathrm{Tr}(M^*M)]^{1/2} = (\sum_{j=1}^{\ell}\sum_{i=1}^{m} \vert m_{ij}\vert^2)^{1/2}$, Frobenius norm of M
$\Vert M\Vert_2$	:=	$\sigma_{max}(M)$, spectral norm of M

For any $W \in \mathbb{F}^{m\times\ell}$, we define the weighted Frobenius norm by weighting elementwise:

$\Vert M\Vert_W$:= $(\sum_{j=1}^{\ell}\sum_{i=1}^{m} |w_{ij}|\,|m_{ij}|^2)^{1/2}$

For any other matrix $N \in \mathbb{F}^{r\times s}$

$M \otimes N$:= $\begin{bmatrix} m_{11}N & \dots & m_{1m}N \\ \dots & \dots & \dots \\ m_{\ell 1}N & \dots & m_{\ell m}N \end{bmatrix}$, the Kronecker product of M and N

I_m := m×m unit matrix

$\mathbb{1}_m$:= m×m matrix filled with 1's

Let $u \in \mathbb{F}^\ell$ where $\mathbb{F} = \mathbb{R}$ or $\mathbb{C}$, then:

$\Vert u\Vert$:= $(u^*u)^{1/2} = (\sum_{i=1}^{\ell} |u_{ij}|^2)^{1/2}$, the Euclidean vector norm

Let $W = P^*P \in \mathbb{C}^{\ell\times\ell}$ be a hermitian, positive definite (weighting) matrix,

$\Vert u\Vert_W$:= $\Vert Pu\Vert = (u^*Wu)^{1/2}$, the weighted Euclidean vector norm

Let columns of $V \in \mathbb{F}^{t\times\ell}$ $(t < \ell)$ be a basis of a subspace of $\mathbb{F}^\ell$, then:

$V^\perp$ $\in \mathbb{F}^{(\ell-t)\times\ell}$ and its columns form a basis for the orthogonal complement of $\mathcal{R}(V)$

P_V := $VV^\dagger$, the orthogonal projector onto $\mathcal{R}(V)$

$P_V^\perp$:= $I - P_V = V^\perp V^{\perp\dagger}$, the orthogonal projector onto $\mathcal{R}(V^\perp)$

For $p(s) \in \mathbb{R}[s]$, $P(s) \in \mathbb{R}[s]^{m\times\ell}$

$\deg p(s)$	:= degree of the polynomial $p(s)$
$\deg[\text{row}_i(P(s))]$	:= max degree of the polynomials in the ith row of $P(s)$
$\text{diag}(d_i)_{i=1}^n$	:= n×n diagonal matrix with $d_1,\ldots,d_n$ along the diagonal; also written as $\text{diag}(d_1,\ldots,d_n)$ or $\text{diag}(d_i)$
$\text{p-diag}(d_i)_{i=1}^n$	:= pseudo-diagonal matrix with $d_1,\ldots,d_n$ along its principal diagonal

Let $\Omega \subset \mathbb{C}$, $f(s) \in \mathbb{R}(s)$ and $G(s) \in \mathbb{R}(s)^{m\times\ell}$, then:

$\#Z(f(s),\Omega)$	:= number of zeros (multiplicities counted) of $f(s)$ in Ω
$\#P(f(s),\Omega)$	:= number of poles (multiplicities counted) of $f(s)$ in Ω
$\#SMZ(G(s),\Omega)$	:= number of Smith-McMillan zeros of $G(s)$ in Ω
$\#SMP(G(s),\Omega)$	:= number of Smith-McMillan poles of $G(s)$ in Ω
$\#IZ(G(s))$	:= number of ∞ zeros (multiplicities counted) of $G(s)$

Let γ be a (finite number of) closed curve(s) in $\mathbb{C}$, then:

$\#E(\gamma,a)$	:= number of encirclements of γ around the point a (our convention is positive for anticlockwise)

D_{NYQ}	Nyquist D-contour, see §1.2
$MS(\cdot)$	measure of skewness, see §1.6
$GL(n,\mathbb{C})$	general linear group, see §2.1
$U(n)$	unitary group, see §2.1
$SU(n)$	special unitary group, see §2.1
$m(G)$	frame misalignment of G, see §3.1
$TPC(G(s))$	total phase change of the characteristic gain loci of $G(s)$, see §4.6

$f \circ g$	denotes the composition of two functions, f after g
$\forall$, $\exists$	denotes for all, there exist(s)
$\square$	marks the end or the absence of a proof

List of Abbreviations:

AIRC	Aircraft Dynamics Model, Appendix F
AUTM	Automobile Gas Turbine Model, Appendix C
CAD	Computer-Aided-Design, §4.1
CGL, CGLi	Characteristic Gain Loci, ith branch of, §1.2
CVD	Characteristic Value Decomposition
CLTM	Closed-Loop Transfer Matrix, §4.3.2
GMI	Gain Margin Interval, §4.4
LHP	Left Half Plane
LQR	Linear Quadratic Regulator
MFD	Matrix Fraction Description
NSRE	Non-Square Chemical Reactor Model, Appendix D
PD	Polar Decomposition, §1.5
PGD	Parameter Group Decomposition, §2.1
PGL, PGLi	Parameter Group Loci, ith branch of, §2.3
PI	Proportional plus Integral
PMI	Phase Margin Interval, §4.4
QND	Quasi-Nyquist Decomposition, §3.2
QNL, QNLi	Quasi-Nyquist Loci, ith branch of, §3.5
REAC	Chemical Reactor Model, Appendix D
RFN	Reversed-Frame Normalizing/Normalization
RFNDP	Reversed-Frame Normalizing Design Procedure, §5.1
RHP	Right Half Plane
SVD	Singular Value Decomposition, §1.3
STD	Schur Triangular Decomposition, §1.6
s.t.	such that
TGEN	Turbo-Generator Model, Appendix E
w.r.t.	with respect to

CHAPTER 0 INTRODUCTION

The purpose of the work presented here is the development of a computer-aided analysis-design approach to linear multivariable feedback systems having the following attributes.

(i) The essence of the classical frequency-response approach to feedback systems is retained as far as possible ; that is one seeks to achieve desired stability and performance targets by the manipulation of gains and phases.

(ii) Performance, stability and robustness are all related to a gain/phase decomposition. For reasons explained below this is based on a systematic use of singular values.

(iii) Controllers are synthesized using least-squares techniques to generate an approximation to an "ideal" controller; hence all the detailed parameter adjustment and tuning is carried out by the computer, following the designer's specification of his requirements.

(iv) Plants having different numbers of inputs and outputs can be handled in a natural way.

In specifying a required feedback control system behaviour the designer will be principally concerned with three aspects of his specification : Stability, Performance and Robustness. By robustness is meant the ability to maintain some specified degree of stability and performance in the face of a stipulated amount of plant variation. Stability can be handled using the generalized Nyquist stability criterion and the associated generalized root locus method [MAC1] [POS1]. However, although generalized Nyquist diagrams give precise information about closed-loop stability, they do not give an adequate characterization of closed-loop performance. This is because the eigenvalues do not give a good description of the gain behaviour of an operator, unless the eigenvectors happen to be an orthogonal set. For example the matrix transfer function

$$G(s) = \begin{bmatrix} 0 & 0 \\ \frac{10^{27}}{(s+1)} & 0 \end{bmatrix}$$

has characteristic gains [MAC1][POS1] which are both identically zero for all values of s, yet it obviously has very large gains for certain inputs. For this reason, characteristic decompositions of an operator are not well suited to the consideration of the performance of feedback systems, and another form of operator decomposition is needed which is more appropriate to the accurate discussion of gain behaviour. Such a decomposition is found in terms of the singular values of an operator, and this has naturally led to an important role for singular value decompositions in feedback systems analysis and design [MAC3][DOY]. The usefulness of singular values is further enhanced by their key role in characterizing robustness [DOY].

The approach to linear feedback control systems developed here is based on the properties of, and the relationships between, three forms of operator decomposition: the singular-value decomposition (SVD), the characteristic-value decomposition (CVD), and the polar decomposition (PD). All three decompositions exist for operators corresponding to systems with the same number of inputs and outputs, while more general operators have only singular-value and polar decompositions. A careful study of the links between the various forms of decomposition in the square case, where the number of inputs and outputs is the same, enables one to relate the properties of feedback loops being formed, which are necessarily associated with "square" operators, to general (non-square) plant descriptions. The relationship between the three decompositions takes a particularly straightforward form when the operator is normal, that is when it has an orthogonal eigenvector framework. An operator which is not normal will be said to be skew, and it is shown that skewness has certain undesirable implications for feedback system behaviour; in particular skewness combined with poor stability margins aggravates a lack of robustness. Hence an approximation to normality is something which one strives to achieve in the feedback design process. Thus normal systems and their properties play a key role in the formulation and implementation of what we call the quasi-classical approach to feedback systems.

The feedback control design problem is considered here in the context of computer-aided design (CAD) using interactive graphic terminals, and appropriate programs have been developed and tested for the techniques described. A designer needs a conceptual framework within which to carry through the complex engineering decisions with which he is faced. Any powerful interactive design technique must present the designer with the full set of indicators required to specify his needs and interpret his results in the context of his conceptual framework. It is the ability to think of the overall aspects of his design in terms of gain and phase parameters, and to associate them with appropriate graphical displays, which makes the quasi-classical approach an attractive one for computer-aided design. The computer is used for calculation, manipulation and optimization. In any fully-developed interactive design technique, the "tuning" of controller parameters is best done by a systematic use of appropriate optimization methods; least-squares fitting techniques play a key role in the ways developed here for generating controller parameters to meet design specifications.

An outline of this monograph is as follows. The next chapter gives some basic definitions and results and deals with singular-value decompositions, polar decompositions, characteristic gain loci, and the generalized Nyquist stability criterion. Special attention is paid to the continuity of singular values and singular vectors, since we wish to study their variation with frequency for a transfer-function matrix representing the dynamical behaviour of a given plant. Other standard results, particularly solutions to least-squares problems, will be given as they are needed in later chapters.

Two different types of Nyquist-like loci are discussed in Chapters 2 and 3. The Quasi-Nyquist loci in the complex gain plane developed in Chapter 3 are obtained by transferring phase information from singular vectors to singular values and are intended specifically for use in feedback design (as opposed to analysis). They form the basis of a design technique introduced in Chapter 4. Essentially, their role is to assess the behaviour of a compensated system after completion of the feedback loop through a specific form of controller (the "reversed-frame-normalizing controller").

The idea of a "reversed-frame-normalizing" design technique is introduced in Chapter 4. This aims at simultaneously manipulating the Quasi-Nyquist loci and normalizing the system's eigenvector framework. It is shown there that, for a normal system (that is one having an orthonormal eigenframework), the basic feedback system properties, that is stability, robustness and closed-loop performance, are characterized in a particularly nice way; and moreover one which can be easily and accurately interpreted in terms of a set of characteristic gain loci (generalized Nyquist diagrams).

Chapter 5 gives details of ways of implementing the reversed-frame-normalizing design procedure, using linear least-squares techniques to synthesize a compensator which approximates an "ideal" one. A refinement of this solution using nonlinear least-squares techniques is then given in Chapter 6.

Some illustrative examples and a brief overall discussion of this quasi-classical approach to multivariable feedback system analysis and design are given in Chapter 7. Details of the various plant models used in these studies are given in a series of appendices.

The singular values of a matrix-valued function of a complex variable, and their relationship to the characteristic values (eigenvalues), play a key role in the work presented here. Since it is known that the characteristic values are locally a set of holomorphic functions which can be globally organized into a Riemann surface, it is a matter of considerable interest to investigate the analytic nature of the singular values. This is briefly discussed in Appendix A, where it is shown that the singular values

$$\{ \sigma_i(x,y) : i=1,2,\ldots,\min(m,\ell) \}$$

of an $m\times\ell$ rational matrix $G(s)$, with $s=x+jy$, are locally a set of real-analytic functions of x and y. This property is of background interest and is not used in the main text, which can therefore be read independently of the results given in Appendix A.

CHAPTER 1 SINGULAR-VALUE, CHARACTERISTIC-VALUE AND POLAR DECOMPOSITIONS

§1.1 System Description and Feedback Configurations

Let $\mathcal{U}(s)$, $\mathcal{Y}(s)$ be the vector spaces $\mathbb{C}(s)^{\ell}$, $\mathbb{C}(s)^{m}$ over the field $\mathbb{C}(s)$ (see the list of symbols on page iii). In the treatment given here, a linear system, $\mathcal{G}(s)$, is a linear-operator-valued function of a complex variable s such that

$$\mathcal{G}(s) : \mathcal{U}(s) \rightarrow \mathcal{Y}(s)$$

has a rational matrix representation $G(s) \in \mathbb{R}(s)^{m\times \ell}$. We will often, with a mild abuse of language, refer to the linear system $G(s)$. $\mathcal{U}(s)$, $\mathcal{Y}(s)$ will be called the input space and the output space of transform vectors respectively. The complex variable s will be called the frequency. When $s = j\omega$ for $\omega \in \mathbb{R}$, ω will be called the angular frequency.

For $u(s) \in \mathcal{U}(s)$, $y(s) \in \mathcal{Y}(s)$, the equation

$$y(s) = G(s)u(s)$$

will be represented diagrammatically as Fig.1.1.

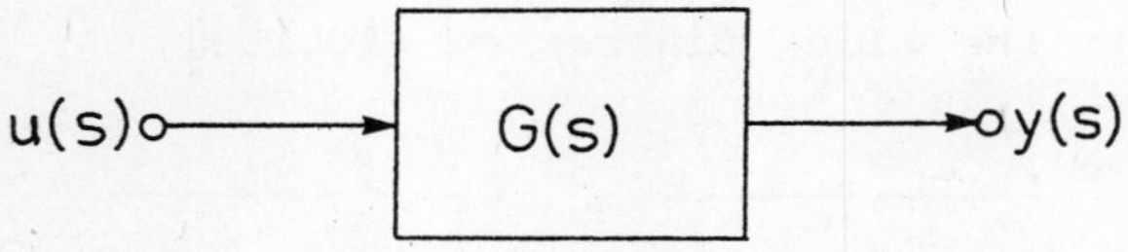

Fig.1.1

Given the linear system $\mathcal{G}(s)$, a compensator $\mathcal{K}(s)$ is a linear system

$$\mathcal{K}(s) : \mathcal{Y}(s) \rightarrow \mathcal{U}(s)$$

with a matrix representation $K(s) \in \mathbb{R}(s)^{\ell \times m}$. The composition of $\mathcal{K}(s)$ and $\mathcal{G}(s)$,

$$\mathcal{G} \circ \mathcal{K}(s) \; : \; \mathcal{Y}(s) \rightarrow \mathcal{Y}(s)$$

is called a precompensated system. $\mathcal{G} \circ \mathcal{K}(s)$ has the matrix representation $G(s)K(s) \in \mathbb{R}(s)^{m \times m}$ and will be represented by the block diagram Fig.1.2.

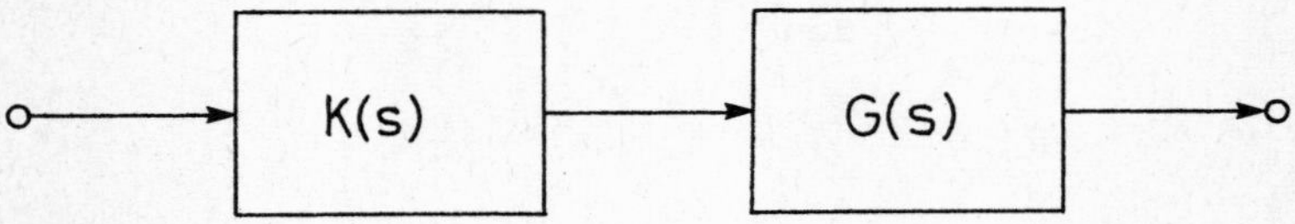

Fig.1.2

For a precompensated system $G(s)K(s)$, the closed-loop system corresponding to a standard negative-unity-feedback configuration is the linear system

$$\mathcal{T}(s) \; : \; \mathcal{Y}(s) \rightarrow \mathcal{Y}(s)$$

represented by

$$T(s) = [I_m + G(s)K(s)]^{-1} G(s)K(s) \quad \in \mathbb{R}(s)^{m \times m}$$

corresponding to the block diagram of Fig.1.3.

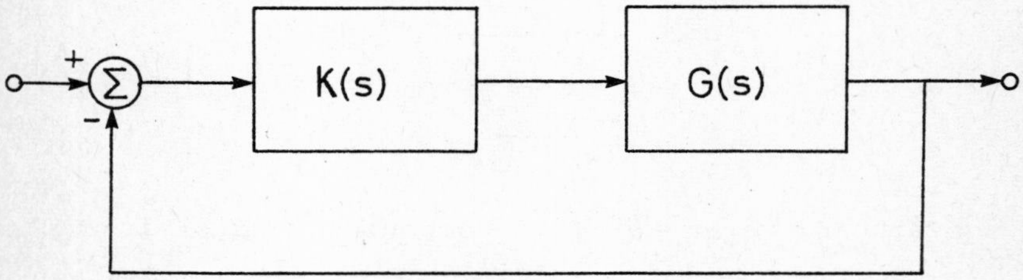

Fig.1.3

A postcompensated system $\mathcal{K} \circ \mathcal{G}(s)$ can be defined in a similar way, in which case the closed-loop system corresponding to the negative-unity-feedback configuration is given by

$$T(s) = \lfloor I_\ell + K(s)G(s) \rfloor^{-1} K(s)G(s) \quad \varepsilon\ \mathbb{R}(s)^{\ell \times \ell}$$

The feedback design (or, more accurately in the restricted context of this investigation, compensator design) problem to be considered is:

> Given a linear system G(s), find a compensator K(s) such that the closed-loop system T(s) satisfies some set of specifications of performance, stability and robustness.

§1.2 Characteristic Gain Loci and the Generalized Nyquist Stability Criterion

Consider a square linear system $G(s) \varepsilon\ \mathbb{R}(s)^{m \times m}$. Let D_{NYQ} denote a parametrization of the usual Nyquist D-contour (see Fig.1.4) with a semi-circular indentation into the left half plane (LHP) if G(s) has poles or zeros on the imaginary axis. The radius R is made sufficiently large to enclose all closed right half plane (RHP) poles (and zeros if necessary). We shall also use the same symbol D_{NYQ} to denote the set of points on the contour. The context will always make the meaning clear.

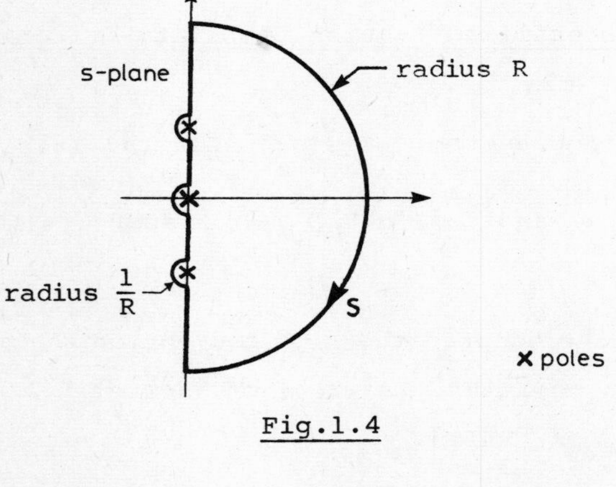

Fig.1.4

As s traverses D_{NYQ} in the clockwise direction, the set of eigenvalues (characteristic values) of G(s)

$$\lambda(G(s)) = \{g_1(s),\ldots,g_m(s)\} \tag{1.2.1}$$

traces out a set of eigenloci in the complex plane. It is well established that the eigenfunctions (1.2.1) may be regarded as branches of a single algebraic function g(s) defined by [MAC1][SMI1] [BLI] (also see Appendix A)

$$\det[g(s)D(s)-N(s)] = 0$$

where $G(s) = N(s)D(s)^{-1}$ is a right coprime matrix fraction description (MFD). Alternatively, by a graph-theoretic approach, it can be shown that the eigenfunctions are differentiable functions of s and that the set of loci (1.2.1) can always be juxtaposed to give a number of closed circuits [DES1]. In either case, we shall denote the combined loci by $g \circ D_{NYQ}$ and the individual branches by $g_i \circ D_{NYQ}$ (i = 1,...,m). They will be called the (ith branch of) <u>characteristic gain loci (CGL)</u> and $g_i \circ D_{NYQ}$ will be labelled as CGLi in diagrams.

If $g \circ D_{NYQ}$ does not pass through the critical point (-1+j0), then it has a well-defined winding number (encirclements), denoted by $\#E(g \circ D_{NYQ},-1)$, around that point. The generalized Nyquist stability criterion can then be stated as

<u>Theorem 1.2.1 (Generalized Nyquist Stability Criterion)</u>

Let $G(s) \in \mathbb{R}_p(s)^{m\times m}$.

Then the closed-loop system $[I+G(s)]^{-1}G(s)$ is stable iff

$$-1 \notin g \circ D_{NYQ} \quad \text{and} \quad \#E(g \circ D_{NYQ},-1) = \#SMP(G(s),\mathbb{C}_+) \qquad \Box$$

Note that we have adopted the sign convention that anticlockwise encirclements are positive. A proof of Theorem 1.2.1 can be found in [MAC1], [DES1] or [SMI1].

As is well-known the benefits of feedback control accrue from deployment of high open-loop gains. In scalar feedback theory the well-known Bode relationships between gain and phase behaviour [BOD] are important because they set the trade-off between the benefits sought from the use of feedback and the consequent price which has to be paid in terms of gain-bandwidth. To get an acceptable compromise between performance and stability for a given amount of available gain-bandwidth, appropriate gain-phase trade-offs must be made. A similar situation arises in the multivariable case. It has been shown that (see [SMI2]) the scalar Bode gain-phase relationships can be extended to the multivariable case, with the role of the scalar transfer function being replaced by the characteristic gain function. For single-input single-output systems, the Bode relations hold only for minimal-phase systems. For multiple-input multiple-output systems an additional restriction, which has no counterpart in the scalar case, has to be made.

Theorem 1.2.2 (Multivariable Bode Gain-Phase Relationships)

Let $G(s) \in \mathbb{R}(s)^{m\times m}$ and suppose that

(1) the polynomial equation defining the characteristic gains of G(s) is irreducible.

(2) G(s) has no poles, zeros or branch points in $\mathbb{C}_+$.

(3) G(0) has real positive eigenvalues.

Then the characteristic gains $g_i(s)$ of G(s) are m separate analytic functions in $\mathbb{C}_+$ each satisfying the Bode relationships

$$\arg g_i(j\omega_o) = \frac{2\omega_o}{\pi}\int_o^\infty \frac{\log|g_i(j\omega)| - \log|g_i(j\omega_o)|}{\omega^2 - \omega_o^2}\, d\omega \qquad (1.2.2)$$

□

A careful, detailed discussion and a proof of this result can be found in [SMI2]. We note that, as far as the assumptions of this theorem are concerned, (1) is generically satisfied and (3) can always be achieved by a suitable modification of G(s) (see the discussion of "standardization at $s=0$" given in Chapter 3). It is clearly necessary to require that $G(s)$ has no poles or zeros in $\mathbb{C}_+$. However, whether it is possible, and if so whether it is necessarily a sensible strategy, to remove branch points in $\mathbb{C}_+$ is a topic which requires further investigation.

§1.3 The Singular Value Decomposition (SVD)

A constant matrix $G \in \mathbb{C}^{m\times \ell}$ represents a linear operator $G : \mathbb{C}^{\ell} \to \mathbb{C}^{m}$ taking a vector $u \in \mathbb{C}^{\ell}$ into the vector $y = Gu \in \mathbb{C}^{m}$. A natural way to look at the gain of the operator G along the direction u is to consider the ratio $\|y\|/\|u\|$ where $\|\cdot\|$ denotes the Euclidean vector norm. This gain, of course, depends on the direction of u. The singular value decomposition can be regarded as a particularly nice way of choosing orthonormal bases in $\mathbb{C}^{\ell}$ and $\mathbb{C}^{m}$ so that the gains of G along the basis vector directions can be characterized by some minimax conditions. A detailed interpretation along such lines, from a systems' point of view, is given in [MAC3]. Many other applications of the singular value decomposition to system theory are given in [KLE]. A review of the relevant results is given below.

Theorem 1.3.1 (Existence of SVD)

Let $G \in \mathbb{C}^{m\times \ell}$, then $\exists$ unitary matrices $Y \in \mathbb{C}^{m\times m}$, $U \in \mathbb{C}^{\ell\times \ell}$ s.t.

$$G = Y\Sigma U^* \qquad (1.3.1)$$

where $\Sigma = \text{p-diag}(\sigma_1,\ldots,\sigma_r) \in \mathbb{R}^{m\times \ell}$, $r = \min(m,\ell)$

and $\sigma_1 \geq \sigma_2 \geq \cdots \geq \sigma_r \geq 0$.

In particular, if $G \in \mathbb{R}^{m\times \ell}$, then Y, U can be chosen to be orthogonal.

□

For a proof, see for example [STE1,pp.318 Theorem 6.1]. Clearly $\sigma_1^2,\ldots,\sigma_r^2$ are the real positive eigenvalues of G^*G (or GG^*) and U, Y are respectively the eigenvector matrices of G^*G, GG^*. The numbers $\sigma_1,\ldots,\sigma_r$ are called the singular values of G and we denote the set

$\{\sigma_1,\ldots,\sigma_r\}$ by $\sigma(G)$. The columns of Y and U are respectively called <u>left and right singular vectors</u> of G.

In the case that $m \neq \ell$, Σ contains a trailing zero block, of order $(m-\ell)\times\ell$ on the bottom if $m > \ell$ or of order $\ell\times(\ell-m)$ on the right if $m < \ell$. This block can simply be deleted, if correspondingly the last $(m-\ell)$ columns of Y (for $m > \ell$) or the last $(\ell-m)$ rows of U^* (for $m < \ell$) are left out. Suppose this is done and the new matrices are given the same names Σ, Y, U^*. Then Σ will be square and either Y or U will consist of part of a unitary matrix, then called a <u>subunitary</u> matrix. For convenience, we shall use the term subunitary as including the case where the matrix is possibly unitary.

We note that the spectral norms of G and G^{-1} (if G is nonsingular, i.e., $m = \ell$ and $\sigma_r > 0$) are given by

$$\|G\|_2 = \sigma_1 \tag{1.3.2}$$

$$\|G^{-1}\|_2 = 1/\sigma_r \tag{1.3.3}$$

The uniqueness of singular values and lack of uniqueness of singular vectors are established in the next proposition.

<u>Prop 1.3.2</u>

In the notation of Theorem 1.3.1:

(1) The singular values $\sigma_1,\ldots,\sigma_r$ are uniquely defined.

(2) If $\sigma_i(\neq 0)$ is distinct from the other singular values, then the corresponding left and right singular vectors y_i, u_i are uniquely defined up to the same scalar factor of the form $e^{j\theta}$.
i.e., if y_i', u_i' are another pair of left and right singular vectors, then $\exists\ e^{j\theta} \in \mathbb{C}$ s.t.

$$y_i' = e^{j\theta}y_i \quad , \quad u_i' = e^{j\theta}u_i \tag{1.3.4}$$

Hence $y_iu_i^*$ and $u_i^*y_i$ are uniquely defined.

<u>Proof</u>:

(1) Since $\sigma_1^2,\ldots,\sigma_r^2$ are the eigenvalues of the hermitian matrices G^*G and GG^*, they are uniquely defined.

(2) Let y_i, u_i be the ith columns of Y, U respectively. From (1.3.1),

$$G^*Gu_i = \sigma_i^2 u_i$$

and $$y_i = \sigma_i^{-1} G u_i \tag{1.3.5}$$

Now u_i, being a normalized unit eigenvector of G^*G corresponding to the eigenvalue σ_i^2, is defined up to a scalar factor of modulus one. So any other right singular vector can be written as

$$u_i' = e^{j\theta} u_i \qquad \text{for some } \theta$$

Once u_i is chosen, the left singular vector y_i is determined by (1.3.5). Suppose u_i' has been taken as the right singular vector instead of u_i, then the corresponding left singular vector would have been

$$y_i' = \sigma_i^{-1} G u_i' = e^{j\theta}(\sigma_i^{-1} G u_i) = e^{j\theta} y_i$$

It follows that $y_i u_i^*$ and $u_i^* y_i$ are uniquely defined. □

Next we consider the case of equal singular values.

<u>Prop 1.3.3</u>

In the notation of Theorem 1.3.1:

(1) If there are t $(1 \leq t \leq r)$ equal singular values, say, $\sigma_i = \dots = \sigma_{i+t-1} (\neq 0)$ with corresponding left and right singular vectors given by the columns of

$$\tilde{Y} = [y_i \ \dots \ y_{i+t-1}] \ \varepsilon \ \mathbb{C}^{m \times t}$$
$$\tilde{U} = [u_i \ \dots \ u_{i+t-1}] \ \varepsilon \ \mathbb{C}^{\ell \times t}$$

then $\tilde{Y}$, $\tilde{U}$ are defined up to postmultiplying by the same $t \times t$ unitary matrix. i.e., if $\tilde{Y}'$, $\tilde{U}'$ are another pair of left and right singular vector matrices, then $\exists$ a unitary matrix $X \ \varepsilon \ \mathbb{C}^{t \times t}$ s.t.

$$\tilde{Y}' = \tilde{Y}X \quad , \quad \tilde{U}' = \tilde{U}X \tag{1.3.6}$$

(2) $\tilde{Y}\tilde{U}^*$ is uniquely defined.

Proof:

(1) It follows from (1.3.1) that

$$G^*G\tilde{U} = \sigma_i^2\tilde{U}$$

and $$\tilde{Y} = \sigma_i^{-1}G\tilde{U}$$

The proof is then similar to that of Prop 1.3.2(2).

(2) This follows immediately from (1.3.6). □

Another way of stating Prop 1.3.3 is that if some singular values are equal, then only the column space of the corresponding singular vectors is well determined. This is because the columns of $\tilde{Y}$, $\tilde{U}$ are orthonormal bases of the subspaces they span and any unitary transformation X within these subspaces produces an alternative choice of orthonormal basis vectors and hence of singular vectors.

§1.4 SVD of a Continuous Matrix-Valued Function

The linear system $G(s) \in \mathbb{R}(s)^{m\times\ell}$ is a matrix-valued function of the complex variable s. If we evaluate G(s) at each $s \in \mathbb{C}$ (in particular, $s \in D_{NYQ}$) and do an SVD for G(s), then the singular values and singular vectors will depend on s. Since G(s) is continuous (in fact analytic) in s, we expect continuity properties to carry over, in a sense to be made clear below, to its SVD. Such continuity properties will be implicitly assumed in Chapters 2 and 3 where we use SVD to define Nyquist-like loci. We shall take the frequency dependent real singular values of G(s) as the gain magnitudes of some gain loci and extract phase information from the singular vectors. Clearly, we would like to have a set of continuous (or at least piecewise continuous) loci. In view of Prop 1.3.2(1) and the next proposition, there is no difficulty with the singular values. But some care is needed for the singular vectors. For notational simplicity, we shall only consider the case $m \geq \ell$.

Prop 1.4.1

Suppose $s_o \in \mathbb{C}$ is not a pole of $G(s) \in \mathbb{R}(s)^{m\times\ell}$ $(m \geq \ell)$ and let $\sigma_1(s) \geq \ldots \geq \sigma_\ell(s) \geq 0$ be the singular values of $G(s)$. Then

(1) $\sigma_i(s)$ $(i = 1,\ldots,\ell)$ are continuous at $s = s_o$.

(2) If $\sigma_i(s_o)$ is distinct from all other singular values of $G(s_o)$, then the corresponding left and right singular vectors $y_i(s)$, $u_i(s)$ can be chosen to be continuous (elementwise) at $s = s_o$.

Proof:

(1) Since $G(s)$ and hence $G^*G(s)$ $(:= G(s)^*G(s))$ are continuous at $s = s_o$, the coefficients of the polynomial equation in $\sigma(s)^2$:

$$\det[\sigma(s)^2 I - G^*G(s)] = 0 \tag{1.4.1}$$

are continuous at $s = s_o$ too. Now the roots of a polynomial equation are continuous functions of its coefficients and it follows that the singular values $\sigma_i(s)$ $(i = 1,\ldots,\ell)$, which are square roots of the roots of (1.4.1), are also continuous at $s = s_o$.

(2) In Prop 1.4.2, we see that if $\sigma_i(s_o)$ is distinct from the other singular values, then the singular vectors are determined except for a phase factor. What we shall show is that we can make a continuous choice of singular vectors which removes this arbitrariness in phase.

First, by taking a sufficiently small neighbourhood N of s_o, we can assume that

$$H(s) := [\sigma_i(s)^2 I - G^*G(s)]$$

has rank $(\ell-1)$ for all $s \in N$. Without loss of generality, we further assume that the first $(\ell-1)$ rows of $H(s)$ are linearly independent. The right null space of $H(s)$ is spanned by

$$h(s) := [H^1(s),\ldots,H^\ell(s)]^T$$

where $H^j(s)$ denotes the cofactor of the (ℓ,j)-entry of $H(s)$. Since $h(s)$ is nonzero $\forall$ $s \in N$, we can take

$$u_i(s) = \frac{h(s)}{\|h(s)\|}$$

to be the right singular vector corresponding to $\sigma_i(s)$. Now the elements of $h(s)$ are polynomial functions of elements of $H(s)$ and since these are continuous at $s = s_o$, so are $h(s)$ and $u_i(s)$.

The left singular vector is now defined by

$$\sigma_i(s)y_i(s) = G(s)u_i(s)$$

and is continuous at $s = s_o$ if $u_i(s)$ is. □

We shall call the frequency dependent $\sigma_i(s)$'s the <u>singular value functions</u>.

Next we consider the case of equal singular values. If at $s = s_o$, $G(s_o)$ has $t(>1)$ equal singular values, then it is well known to the numerical analyst that the singular vectors can be very sensitive to a small perturbation of $G(s_o)$. Stewart has constructed an example [STE2] to show that elements of singular vectors can "jump" by an order of unity for an arbitrary small perturbation of a hermitian matrix. In other words, we cannot expect, in general, the singular vectors to vary continuously at $s = s_o$. Fortunately, the subspace defined by the t singular values does vary continuously. This means that whenever the singular values cluster together, we have to consider subspaces spanned by the singular vectors instead of the individual vectors.

§1.5 Polar Decomposition (PD)

In a singular value decomposition

$$G = Y\Sigma U^* \qquad (1.5.1)$$

we will often refer to the various components by the following convenient names:

Y := Output singular-vector frame matrix

Σ := Principal gain matrix

U := Input singular-vector frame matrix

Any matrix whose columns span some linear vector space, thus forming a basis for that space, will be called a frame matrix. The columns of Y may be called output gain directions, and the columns of U called input gain directions. We will use the terms singular value and principal gain interchangeably, according to context. When discussing system behaviour, we prefer the more physically illuminating term principal gain [MAC3][POS2] and when discussing mathematical or numerical analysis aspects we will use the term singular value. In general if $G \in \mathbb{C}^{m\times \ell}$ then $Y \in \mathbb{C}^{m\times r}$, $U \in \mathbb{C}^{r\times \ell}$ where $r=\min(m,\ell)$ (see discussion after Theorem 1.3.1). Since Y, U are subunitary, we have

$$Y^*Y = I_r = U^*U$$

We may therefore write G in the forms

$$G = (Y\Sigma Y^*)(YU^*) = M_\ell \Phi \tag{1.5.2}$$

$$G = (YU^*)(U\Sigma U^*) = \Phi M_r \tag{1.5.3}$$

where

$$\Phi := YU^* := \text{Phase matrix (of G)}$$

$$M_\ell := Y\Sigma Y^* := \text{Left modulus matrix (of G)}$$

$$M_r := U\Sigma U^* := \text{Right modulus matrix (of G)}$$

It is convenient at this point to also define

$$UY^* := \text{Inverse phase matrix (of G)}$$

$$U^*Y := \text{Alignment matrix (of G)}$$

The alignment matrix is used in Chapter 3; when it is diagonal the input and output gain frames are said to be aligned. It is shown in §3.1 that alignment, in this sense, implies normality.

If G is square, the phase matrix YU^* is a unitary matrix and so has a characteristic decomposition of the form

$$YU^* = \Phi = P\Theta P^* \tag{1.5.4}$$

where P is a unitary matrix and Θ is a diagonal matrix formed from the spectrum of Φ. We will call P the phase frame matrix of G and its columns the phase directions. Since the phase matrix is unitary, Θ will have the form

$$\Theta = \mathrm{diag}(e^{j\theta_i})$$

and we refer to the set of angles θ_i as

$$\theta_i := \text{principal phases of } G$$

This is consistent with the convention

$$\Sigma = \mathrm{diag}(\sigma_i)$$

where $\sigma_i :=$ principal gains of G

Further discussion of principal gains and principal phases may be found in [POS2].

The polar decompositions (1.5.2) and (1.5.3) are the matrix analogues of the polar decompositions of a complex number. It is interesting to note that the polar decomposition is simply a re-packaging of the information contained in the singular value decomposition.

§1.6 Normality and Spectral Sensitivity

A matrix Q is said to be normal if it commutes with its conjugate transpose

$$QQ^* = Q^*Q$$

Examples of normal matrices include orthogonal matrices, unitary matrices and hermitian matrices. The reason why normal matrices are important in our context is brought out by the following equivalent characterization.

Theorem 1.6.1

$Q \in \mathbb{C}^{m\times m}$ is normal iff Q has a complete orthonormal system of eigenvectors, i.e., $\exists$ unitary matrix W s.t. Q has eigenvalue decomposition

$$Q = W\Lambda W^* \tag{1.6.1}$$

where $\Lambda = \mathrm{diag}(\lambda_1,\ldots,\lambda_m)$

Proof:

For example, see [GAN, vol.I pp.272 Theorem 4 and pp.273 Theorem 4']. □

It is well-established in the numerical analysis literature that the eigenvalues of a normal matrix are relatively insensitive to perturbations. In fact, if a normal matrix Q is perturbed to $Q(I+\Delta)$, then it can be shown that any eigenvalue λ of $Q(I+\Delta)$ is bounded within a disc centred around some eigenvalue λ_i of Q (e.g. see [WIL, Chapter 2 §30 and §31]) given by

$$\begin{aligned} |\lambda - \lambda_i| &\leq |\lambda_i|\,\|\Delta\|_2 \\ &\leq \|Q\|_2\,\|\Delta\|_2 \end{aligned} \tag{1.6.2}$$

The insensitivity of the spectrum of a normal matrix is relevant to our investigations in the following way. For the sake of argument, suppose a linear system Q(s) is normal for all $s \in D_{NYQ}$; then its spectrum, and hence its characteristic gain loci will be insensitive to perturbations of Q(s). The importance of the insensitivity of the characteristic gain loci will become apparent when we come to consider the problem of maintaining the stability of a system subjected to perturbations in its plant dynamics.

Although normal matrices have nice spectral properties, they constitute only a relatively small set among general matrices and so the above observations would be more useful if they applied to approximately normal matrices. (For each normal matrix, there is a whole neighbourhood of approximately normal ones.) This is indeed the case. To see this, we introduce a terminology for matrices that depart from normality, and a measure for the departure.

Definition 1.6.2

A matrix is said to be skew iff it is not normal.

Now by a classical result due to Schur, any matrix $Q \in \mathbb{C}^{m\times m}$ can be decomposed as (e.g. see [WIL, Chapter 1 §47])

$$\begin{aligned} Q &= S T_u S^* \\ &= S(D+T)S^* \end{aligned} \tag{1.6.3}$$

where S is unitary, T_u is upper triangular and D, T are respectively the diagonal and strictly upper triangular parts of T_u. We shall refer to (1.6.3) as the Schur triangular decomposition (STD). Clearly, the diagonal elements $\lambda_1,\ldots\lambda_m$ of D are the eigenvalues of Q and Q is normal iff T = 0. Hence it is reasonable to define a measure of skewness of Q to be

$$MS(Q) := \frac{\|T\|}{\|Q\|} \tag{1.6.4}$$

where $\|\cdot\|$ denotes the Frobenius norm (see pp.iv). Note that although an STD is not unique, $\|T\|$ is independent of the particular STD taken because

$$\begin{aligned} \|T\|^2 &= \|Q\|^2 - \|D\|^2 \\ &= \|Q\|^2 - \sum_{i=1}^{m} |\lambda_i|^2 \end{aligned}$$

The variation of the spectrum of a skew matrix Q, when perturbed to $Q(I+\Delta)$, can now be bounded in terms of MS(Q). By a result of [HEN] (also see [WIL, Chapter 3 §50]), for any eigenvalue λ of $Q(I+\Delta)$, there exists an eigenvalue λ_i of Q s.t.

$$\frac{|\lambda - \lambda_i|}{1 + \frac{MS(Q)}{\alpha} + \cdots + \frac{MS(Q)^{m-1}}{\alpha^{m-1}}} \leq \|Q\|_2 \|\Delta\|_2 \tag{1.6.5}$$

where

$$\alpha := \frac{|\lambda - \lambda_i|}{\|Q\|}$$

If Q is normal, then MS(Q) = 0 and (1.6.5) reduces to (1.6.2). The point is, however, that if Q is close to normality in the sense that Q has a small skewness measure MS(Q), then we can still expect its spectrum to remain reasonably insensitive to perturbations of Q because of (1.6.5).

CHAPTER 2 USE OF PARAMETER GROUP DECOMPOSITION TO GENERATE NYQUIST-TYPE LOCI

Our objective is the development of ways of decomposing operators in terms of gains and phases, and then using such decompositions in analysis and design. In this chapter we explore a way of decomposing a matrix G into a product of elementary blocks by means of the parametrization of certain matrix groups. Underlying this decomposition is the idea that one can associate a complex matrix G, regarded as a linear operator, with a set of gains and angles. Applying the decomposition to the frequency-dependent system operator G(s) then gives a set of Nyquist-type loci. The term Nyquist-type is used because, like the classical and generalized Nyquist loci, they are related to gains and phases. This particular set of loci has gains defined by the singular-value functions, and hence they are better indicators of the system's gain behaviour than the characteristic gain loci (generalized Nyquist diagrams). However, unlike the characteristic gain loci, they do not give an accurate assessment of closed-loop stability.

Since the parametric decomposition of the unitary and orthogonal groups in terms of angles is well known, it is natural to seek to decompose the linear operators we are interested in via unitary groups and an appropriate gain-describing component. As well as being a natural starting point, such an investigation has its own intrinsic interest. A detailed study of the 2×2 case however shows that this approach has a major drawback. This essentially arises from the way in which phase information is related to the operator decomposition. In constructing a set of Nyquist-like loci, we would naturally like these loci to always carry the significant and relevant phase information to a large degree. It turns out that, with this form of decomposition, one has in certain cases to use a set of frame angles as well as the phase information contained in the loci when interpreting the significance of the loci. This is clearly unsatisfactory, and it is for this reason that only a limited investigation of these parametric group decompositions is given here. In the following chapter we develop an alternative means of handling phase for Nyquist-type loci which does not suffer from this disadvantage.

§2.1 Some Matrix Groups and their Parametrizations

The matrix groups we shall consider are:

$GL(n,\mathbb{C}) := \{ G \in \mathbb{C}^{n\times n} \mid \det G \neq 0\}$ - general linear group

$U(n) := \{ U \in \mathbb{C}^{n\times n} \mid U^*U = I \}$ - unitary group

$SU(n) := \{ S \in U(n) \mid \det S = 1\}$ - special unitary group

These groups have been very well studied by mathematicians and physicists. A very nice introduction is [CUR]. There is more than one way of parametrizing the above groups and we shall go into detail about the one useful to us. The parametrization given below is adapted from standard work (e.g. see [MUR, chapter 2]) on this topic; however we have made some modifications to suit our purpose.

Lemma 2.1.1

Any $S \in SU(2)$ is of the form

$$S = \begin{bmatrix} a & -\bar{b} \\ b & \bar{a} \end{bmatrix}$$

with

$$a\bar{a} + b\bar{b} = 1 \ ; \quad a,b \in \mathbb{C}$$

Proof:

Let

$$S = \begin{bmatrix} a & c \\ b & d \end{bmatrix} \quad \in SU(2)$$

By definition, $ad - bc = 1$ (2.1.1)

and

$$S^*S = \begin{bmatrix} \bar{a} & \bar{b} \\ \bar{c} & \bar{d} \end{bmatrix} \begin{bmatrix} a & c \\ b & d \end{bmatrix} = I$$

from which $\bar{a}a + \bar{b}b = 1$ (2.1.2a)

$\bar{a}c + \bar{b}d = 0$ (2.1.2b)

From (2.1.2b)

$$\frac{d}{\bar{a}} = -\frac{c}{\bar{b}} = k \qquad \text{for some } k$$

Substituting $d = k\bar{a}$ and $c = -k\bar{b}$ into (2.1.1) gives

$$k(a\bar{a} + b\bar{b}) = 1$$

It follows from (2.1.2a) that $k = 1$ and hence $d = \bar{a}$, $c = -\bar{b}$. □

Prop 2.1.2 (Parametrization of SU(2))

Any $S \in SU(2)$ can be written as:

$$S = \begin{bmatrix} \cos\phi & -e^{-j\delta}\sin\phi \\ e^{j\delta}\sin\phi & \cos\phi \end{bmatrix} \begin{bmatrix} e^{j\theta} & 0 \\ 0 & e^{-j\theta} \end{bmatrix}$$

with $-\pi/2 \leq \phi \leq \pi/2$ $\qquad -\pi < \theta \leq \pi$

$-\pi/2 < \delta \leq \pi/2$

Proof:

Write S in the form of Lemma 2.1.1 and let

$$a = |a|e^{j\theta} \qquad -\pi < \theta \leq \pi \tag{2.1.3}$$

Then

$$S = \begin{bmatrix} |a|e^{j\theta} & -\bar{b} \\ b & |a|e^{-j\theta} \end{bmatrix}$$

$$= \begin{bmatrix} |a| & -\bar{z} \\ z & |a| \end{bmatrix} \begin{bmatrix} e^{j\theta} & 0 \\ 0 & e^{-j\theta} \end{bmatrix} \tag{2.1.4}$$

where $z = be^{-j\theta}$. Since $|z| = |b| \leq 1$, we can put

$$z = e^{j\delta}\sin\phi \tag{2.1.5}$$

for some $-\pi/2 \leq \phi \leq \pi/2$, $-\pi/2 < \delta \leq \pi/2$. It follows from (2.1.2a) that

$$|a| = \sqrt{1-|b|^2} = \sqrt{1-|z|^2}$$
$$= \sqrt{1-\sin^2\phi} = \cos\phi \qquad (2.1.6)$$

Putting (2.1.5) and (2.1.6) into (2.1.4) then gives the required result. □

Note that the angle θ is arbitrary when $a = 0$ (see (2.1.3)) and δ is arbitrary when $z = 0$ (see (2.1.5)). These correspond to cases when S has zero diagonal or zero off-diagonal entries. Apart from such cases, the parametrization is unique because of the restrictions of the angles to the appropriate intervals. Alternatively, we may remove these restrictions but then identify the pair of angles (ϕ,δ) with $(-\phi,\delta+\pi)$. Of course the addition of multiples of 2π to any of ϕ, δ, θ makes no difference to S either.

Using Prop 2.1.2, we can now parametrize a 2×2 unitary matrix.

Prop 2.1.3 (Parametrization of U(2))

Any $U \in U(2)$ can be written as:

$$U = \begin{bmatrix} \cos\phi & -e^{-j\delta}\sin\phi \\ e^{j\delta}\sin\phi & \cos\phi \end{bmatrix} \begin{bmatrix} e^{j\theta_1} & 0 \\ 0 & e^{j\theta_2} \end{bmatrix}$$

with $-\pi/2 \leqslant \phi \leqslant \pi/2$ $\qquad -\pi < \theta_1, \theta_2 \leqslant \pi$

$-\pi/2 < \delta \leqslant \pi/2$

Proof:

Let $U \in U(2)$ and let $\det U = e^{j\beta}$.

Then $e^{-j\beta/2}U$ has determinant 1 and so is in SU(2). By Prop 2.1.2, we can write

$$e^{-j\beta/2}U = \begin{bmatrix} \cos\phi & -e^{-j\delta}\sin\phi \\ e^{j\delta}\sin\phi & \cos\phi \end{bmatrix} \begin{bmatrix} e^{j\theta} & 0 \\ 0 & e^{-j\theta} \end{bmatrix}$$

Multiplying both sides by $e^{j\beta/2}$ and setting $\theta_1 = \theta+\beta/2$, $\theta_2 = \theta-\beta/2$ then gives the required result. □

Clearly, the parametrization is unique except when U has a zero diagonal or zero off-diagonal.

In order to parametrize a general 2×2 complex matrix G, we have to use an SVD and apply the last proposition to the relevant unitary matrices of singular vectors. The procedure required is established in the proof of the next theorem.

Theorem 2.1.4 (Parametrization of $GL(2,\mathbb{C})$)

Any $G \in GL(2,\mathbb{C})$ can be written as:

$$G = \begin{bmatrix} \cos\phi_y & -e^{-j\delta_y}\sin\phi_y \\ e^{j\delta_y}\sin\phi_y & \cos\phi_y \end{bmatrix} \begin{bmatrix} \sigma_1 e^{j\theta_1} & 0 \\ 0 & \sigma_2 e^{j\theta_2} \end{bmatrix} \begin{bmatrix} \cos\phi_u & -e^{-j\delta_u}\sin\phi_u \\ e^{j\delta_u}\sin\phi_u & \cos\phi_u \end{bmatrix}^*$$

with $\quad -\pi/2 \leq \phi_y, \phi_u \leq \pi/2 \qquad -\pi < \theta_1, \theta_2 \leq \pi$

$\quad -\pi/2 < \delta_y, \delta_u \leq \pi/2 \qquad 0 < \sigma_2 \leq \sigma_1$

Proof:

Let G have an SVD

$$G = Y \Sigma U^* \tag{2.1.7}$$

where $\Sigma = \mathrm{diag}(\sigma_1, \sigma_2)$ and $Y, U \in U(2)$.

Y, U* can be parametrized as

$$Y = \begin{bmatrix} \cos\phi_y & -e^{-j\delta_y}\sin\phi_y \\ e^{j\delta_y}\sin\phi_y & \cos\phi_y \end{bmatrix} \begin{bmatrix} e^{j\theta_{y1}} & 0 \\ 0 & e^{j\theta_{y2}} \end{bmatrix} \tag{2.1.8a}$$

$$U^* = \begin{bmatrix} e^{-j\theta_{u1}} & 0 \\ 0 & e^{-j\theta_{u2}} \end{bmatrix} \begin{bmatrix} \cos\phi_u & -e^{-j\delta_u}\sin\phi_u \\ e^{j\delta_u}\sin\phi_u & \cos\phi_u \end{bmatrix}^* \tag{2.1.8b}$$

with the angles lying in appropriate intervals. Substituting (2.1.8a,b) into (2.1.7) and setting $\theta_i = \theta_{yi} - \theta_{ui}$ $(i = 1,2)$ (modulo 2π if θ_1, θ_2 get out of the range $(-\pi,\pi]$), we get the stated parametrization of G. This parametrization is unique except when

(1) Y or U have zero diagonal or zero off-diagonal entries, in which case there are non-unique choices of angles in (2.1.8a) or (2.1.8b);

(2) $\sigma_1 = \sigma_2$, in which case Y, U of (2.1.7) may be replaced by YX, UX for any $X \in U(2)$ (see Prop 1.3.3).

Apart from these cases, the parametrization is also independent of the SVD (2.1.7) chosen because of Prop 1.3.2(2). □

We shall refer to the decomposition of G in Theorem 2.1.4 as the parameter group decomposition (PGD).

§2.2 Dimension of Matrix Groups

The matrix groups $GL(n,\mathbb{C})$, $U(n)$ and $SU(n)$ can all be regarded as manifolds having the following dimensions (e.g. see [CUR, pp.39 & pp.103]):

$$\dim GL(n,\mathbb{C}) = 2n^2$$
$$\dim U(n) = n^2$$
$$\dim SU(n) = n^2 - 1$$

For $n = 2$, the above dimensions are 8, 4 and 3. These are, of course, the number of free parameters used for parametrizing $GL(2,\mathbb{C})$, $U(2)$, $SU(2)$ respectively.

The idea of dimension of a matrix group is not needed in any subsequent discussion. However, it is useful, at least in a heuristic sense, to be aware of the number of parameters we are dealing with.

In passing, we mention that, given an n-input n-output system G(s) which is nonsingular for all s lying on a curve $\mathscr{C}$ in $\mathbb{C}$, we may regard the matrix representation of $\mathscr{G} \circ \mathscr{C}$ as the locus of a curve in $GL(n,\mathbb{C})$. Taking such a view, the precompensator design problem can then be looked upon as: how to find a K(s) which will manipulate the curve $\mathscr{G} \circ \mathscr{C}$ into a more desirable curve $(\mathscr{G} \circ \mathscr{K}) \circ \mathscr{C}$ in $GL(n,\mathbb{C})$. This is nothing more than re-phasing the problem in a more abstract language.

§2.3 Nyquist-Type Loci — the PG Loci

Consider a 2-input 2-output system having a transfer function matrix $G(s) \varepsilon \mathbb{R}_p(s)^{2\times 2}$. If $G(s)$ has neither poles nor zeros on the Nyquist D-contour D_{NYQ}, then the image curve $G \circ D_{NYQ}$ lies in $GL(2,\mathbb{C})$ and we can apply the parametrization of $GL(2,\mathbb{C})$ to each point of $G \circ D_{NYQ}$. Specifically, let $s \varepsilon D_{NYQ}$ and let $G(s)$ have a PGD

$$G(s) = Y(s)\Gamma(s)U(s)^* \tag{2.3.1}$$

where $$\Gamma(s) = \mathrm{diag}\big(\gamma_1(s), \gamma_2(s)\big) \tag{2.3.2}$$

$$Y(s) = \begin{bmatrix} \cos\phi_y(s) & -e^{-j\delta_y(s)}\sin\phi_y(s) \\ e^{j\delta_y(s)}\sin\phi_y(s) & \cos\phi_y(s) \end{bmatrix} \varepsilon\, SU(2) \tag{2.3.3a}$$

$$U(s) = \begin{bmatrix} \cos\phi_u(s) & -e^{-j\delta_u(s)}\sin\phi_u(s) \\ e^{j\delta_u(s)}\sin\phi_u(s) & \cos\phi_u(s) \end{bmatrix} \varepsilon\, SU(2) \tag{2.3.3b}$$

The angles $\phi_y(s)$, $\delta_y(s)$ and $\phi_u(s)$, $\delta_u(s)$ will be called the <u>output frame angles</u> and the <u>input frame angles</u> respectively.

Now as s traverses D_{NYQ} in the standard (clockwise) direction, $\gamma_1(s)$, $\gamma_2(s)$ trace out loci on the complex plane. We shall refer to this set of loci, obtained by parameter group decomposition, simply as <u>PG loci</u>. In the diagrams to follow, the loci of $\gamma_1(s)$, $\gamma_2(s)$ will be labelled by PGL1, PGL2 respectively. Similarly, we can plot the input and output frame angles as a function of the frequency $s = j\omega \varepsilon D_{NYQ}$. Before going on to the properties of these loci, we give some examples.

<u>Example 2.3.1</u> (PG loci of Gas Turbine: AUTM)

The system $G(s)$ considered is a 2-input, 12-state, 2-output gas turbine model which will be referred to as AUTM. This is a modified version of the one studied in [EDM2]. Details of AUTM are listed in Appendix C. Fig.2.1(a) shows the PG loci plotted on the complex plane (Nyquist-type diagram) and Fig.2.1(b) shows magnitude and phase versus angular frequency plots (Bode-type diagrams). Fig.2.1(c),(d) show the frame angles versus frequency. Note that when plotting the frame angles, we have removed the restrictions that these angles should lie in the appropriate intervals and instead allow the identifications mentioned after Prop 2.1.2, i.e.,

(ϕ_y, δ_y) with $(-\phi_y, \delta_y+\pi)$

(ϕ_u, δ_u) with $(-\phi_u, \delta_u+\pi)$

any angle with $(2k\pi$ + the same angle), k an integer

The purpose of this is to avoid jump discontinuities when the angles reach their interval boundaries, which tends to obscure the diagrams.

Example 2.3.2

Let

$$G(s) = \begin{bmatrix} \frac{1}{s+1} & 0 \\ \frac{1}{s^2+0.4s+4.04} & \frac{1}{s+3} \end{bmatrix}$$

The corresponding PG loci and frame angles plots are given in Fig.2.2(a-d).

This system has four poles at $\{-1, -3, -0.2\pm 2j\}$. The pair of resonant modes $-0.2\pm 2j$ causes the magnitude of PGL1 to peak, PGL2 to dip, and the frame angles to vary rapidly near the frequency $s = 2j$.

We note that because G(s) is lower triangular, the characteristic gain (eigenvalue) loci of G(s) are obtained from $g_1 = 1/(s+1)$, $g_2 = 1/(s+3)$ (see Fig.2.2(e,f)) and thus miss the resonant modes completely. This example is chosen to exaggerate the differences between the characteristic gain loci and the PG loci.

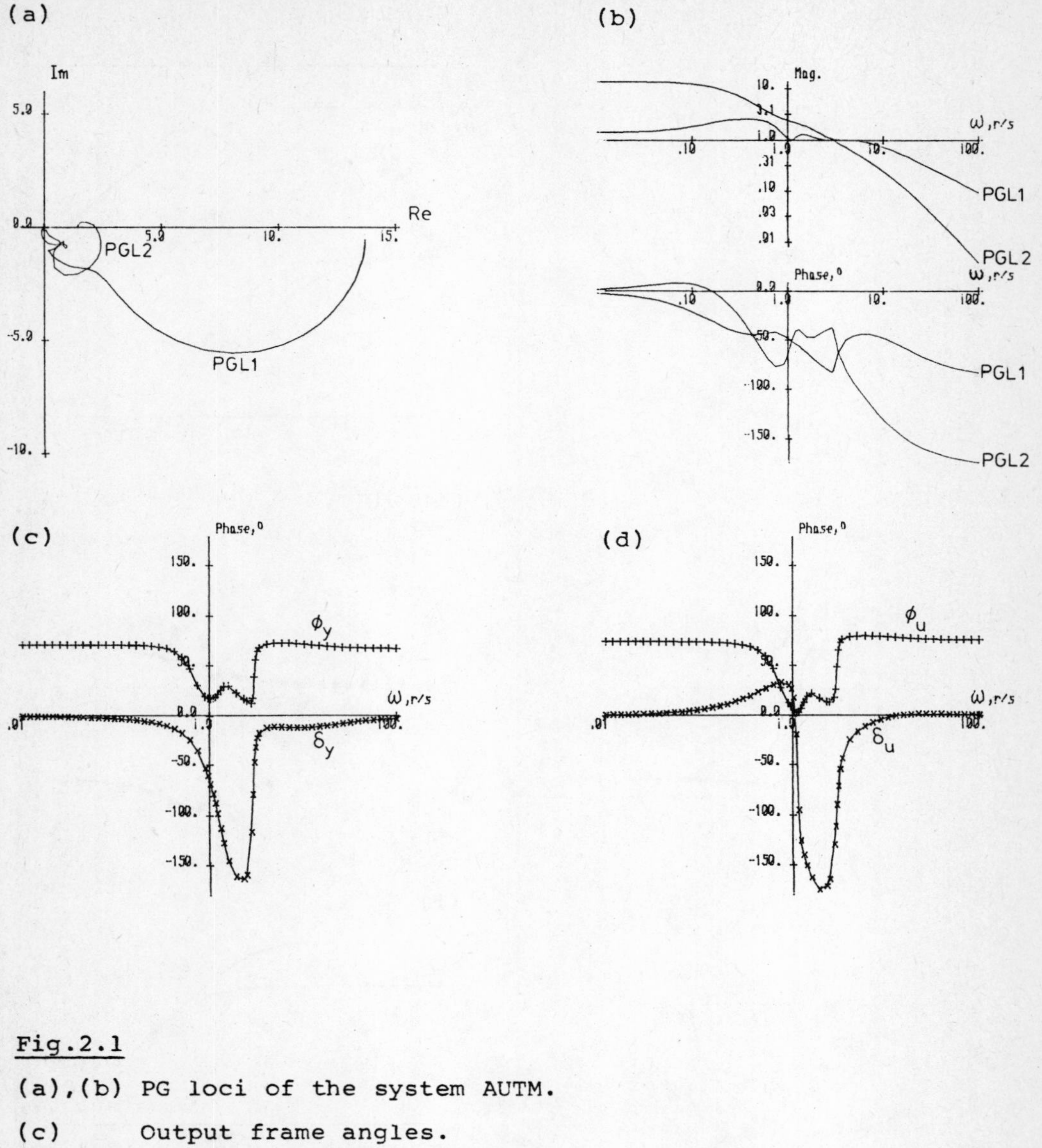

<u>Fig.2.1</u>

(a),(b) PG loci of the system AUTM.

(c) Output frame angles.

(d) Input frame angles.

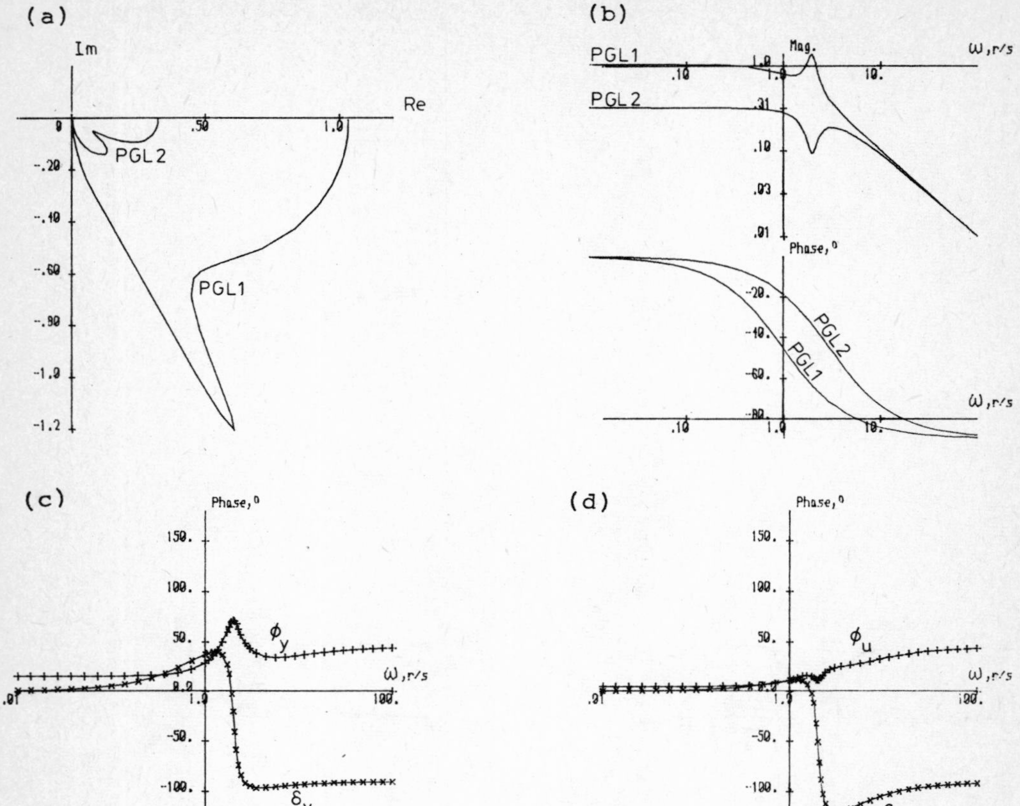

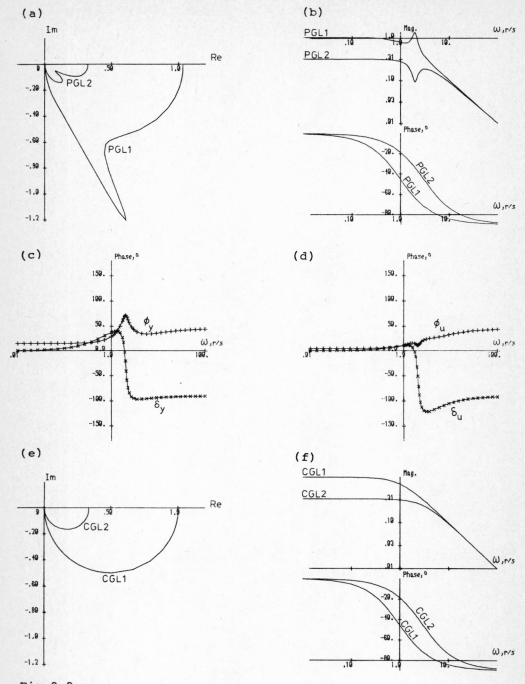

Fig.2.2

(a),(b) PG loci of a system with off-diagonal resonant modes.

(c),(d) Output and input frame angles.

(e),(f) Characteristic gain loci.

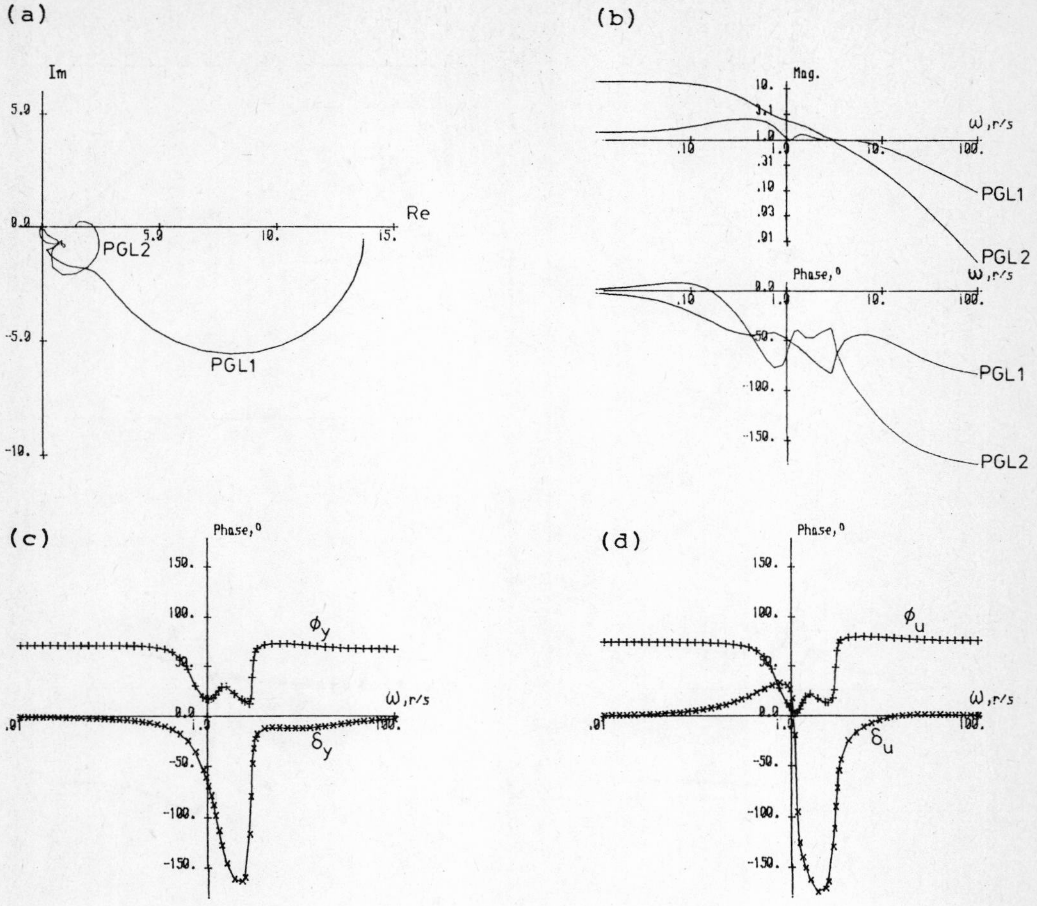

<u>Fig.2.1</u>

(a),(b) PG loci of the system AUTM.

(c) Output frame angles.

(d) Input frame angles.

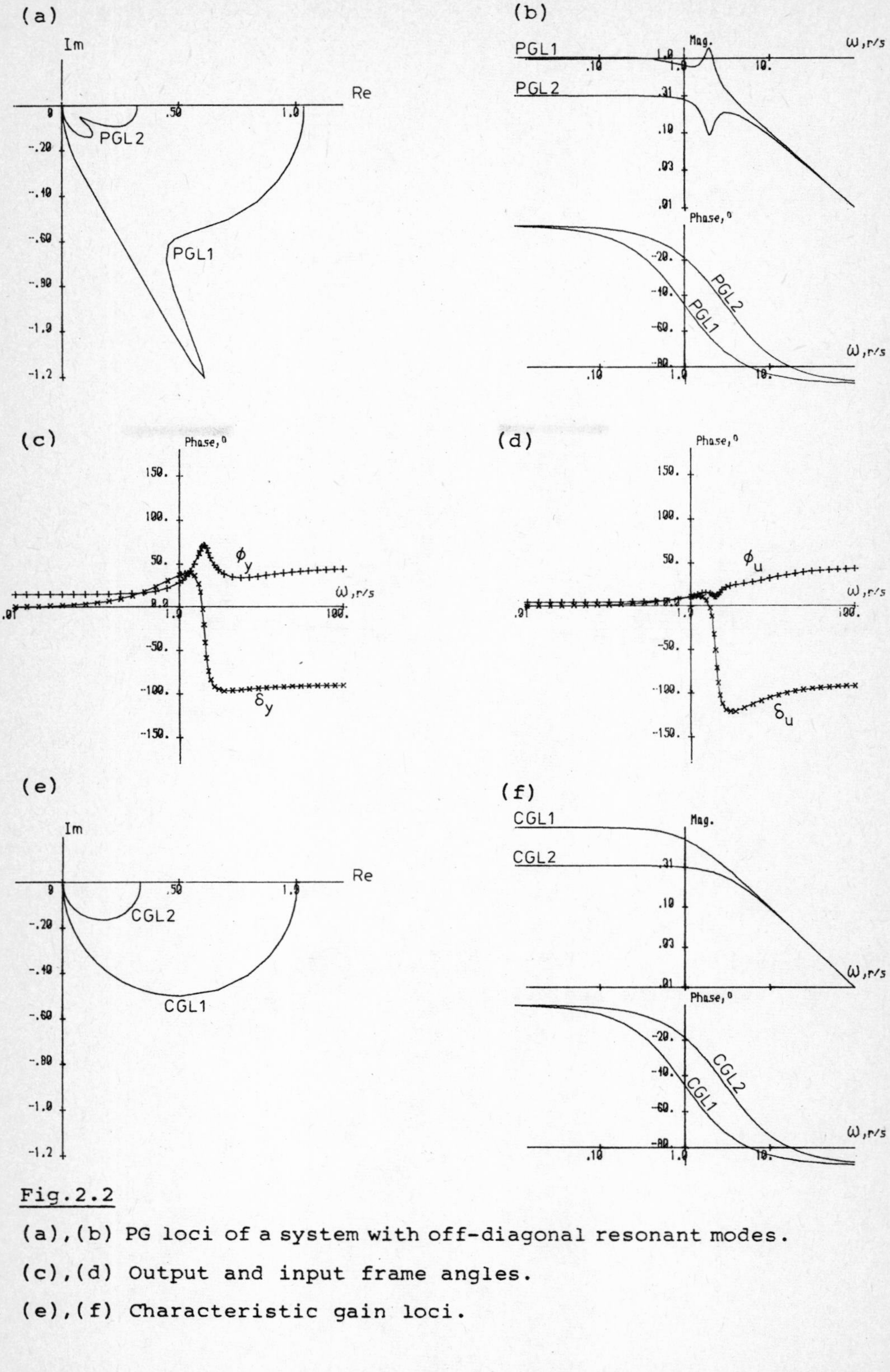

Fig.2.2

(a),(b) PG loci of a system with off-diagonal resonant modes.

(c),(d) Output and input frame angles.

(e),(f) Characteristic gain loci.

A few properties of the frame angles and PG loci should now be obvious from the examples:

(1) At dc, i.e. $s = 0$, $G(0)$ is real and so $Y(0)$, $\Gamma(0)$, $U(0)$ of (2.3.1) are all real. This means that the PG loci always start from points on the real axis and the frame angles $\delta_y(0)$, $\delta_u(0)$ always start from 0°.

(2) If $G(s) \in \mathbb{R}_{sp}(s)^{2\times 2}$, as was the case for the last two examples, then it can be shown that asymptotically, as $s = j\omega \to j\infty$, the PGD of $G(s)$ takes the form (e.g. see [HUN])

$$G(s) \simeq Y(s) \begin{bmatrix} \frac{\sigma_1}{s^{i_1}} & 0 \\ 0 & \frac{\sigma_2}{s^{i_2}} \end{bmatrix} U(s)^*$$

where σ_1, σ_2 are real constants (possibly negative) and i_1, i_2 are the orders of the two infinite zeros of $G(s)$. Moreover, as $s \to j\infty$, $Y(s)$, $U(s)$ approach orthogonal matrices. Thus the asymptotic behaviour of the group parameters is given by:

(i) $\gamma_1(s) \simeq \frac{\sigma_1}{s^{i_1}}$ $\qquad \gamma_2(s) \simeq \frac{\sigma_2}{s^{i_2}}$

which implies that the PG loci have magnitudes rolling off at $-20i_1$ and $-20i_2$ db/decade and phases approaching multiples of $\frac{\pi}{2}$.

(ii) $\delta_y(s)$, $\delta_u(s)$ approach 0 or $\pm\pi$.

For example, judging from the roll-off rates of the PG loci of AUTM (see Fig.2.1(b)), we can deduce that the two infinite zeros are of orders 1 and 2 respectively.

(3) By the results of Chapter 1 about the continuity of singular values and the almost-always continuity of singular vectors, one may deduce that the PG loci are almost-always piecewise continuous. It is true that the PG loci may become undefined at points where the parametrization in Theorem 2.1.4 is not unique; however, we can define the PG loci at such points by continuity arguments. But such results are of no great consequence and we shall thus not go into unnecessary detail.

§2.4 Relationship between the Parameter Group Decomposition and Normality

The normality of a matrix, or rather approximate normality of $G(j\omega)$ over a band of frequencies, will be a recurrent theme throughout this monograph. The reason for this will become clear in Chapters 4 and 5. In this section, we illustrate by an example what we mean by approximate normality over a band of frequencies. First, we state the following obvious fact.

Prop 2.4.1

Let $G \in GL(2,\mathbb{C})$ have a PGD given by Theorem 2.1.4. If the input frame angles are equal to the output frame angles, i.e.,

$$\phi_y = \phi_u \ , \qquad \delta_y = \delta_u \tag{2.4.1}$$

then G is normal. Moreover, the PGD is then an eigenvalue decomposition of G. □

This implies that if at some s_o, the 2 sets of frame angles coincide, then the PG loci will intersect the characteristic gain loci at that frequency. Loosely speaking, if (2.4.1) is approximately satisfied, then we would expect G to be close to a normal matrix and the PG loci to come close to the characteristic gain loci.

Example 2.4.2

Let

$$G(s) = \begin{bmatrix} \frac{1}{s+1} & \frac{2}{s+3} \\ \frac{1}{s+1} & \frac{1}{s+1} \end{bmatrix}$$

This example is taken from [ROS, pp.122 Example 2.1]. The PG loci and frame angles are given in Fig.2.3(a-e). The closeness of the two sets of frame angles says that the PG loci should approximately match the characteristic gain loci (see Fig.2.3(d,e)) over the whole frequency range.

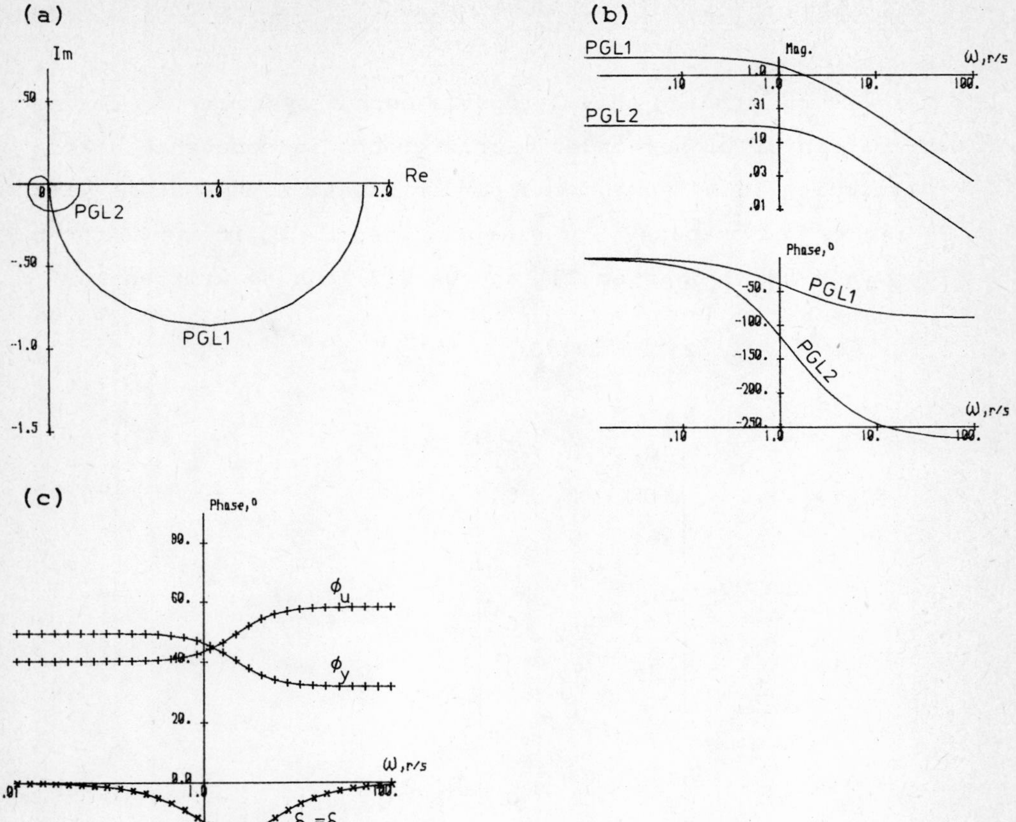

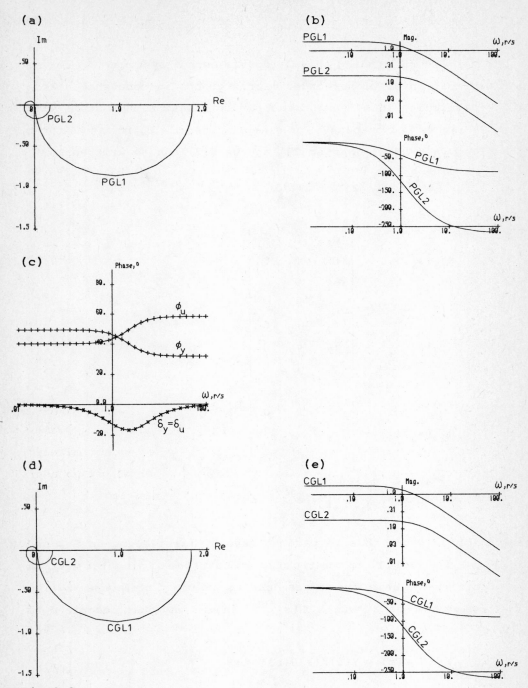

Fig.2.3

(a),(b) PG loci of an approximately normal system.

(c) Output and input frame angles.

(d),(e) Characteristic gain loci.

§2.5 Parametrization of Higher Order Matrix Groups

So far we have studied only 2-input 2-output systems. As far as parametrization of higher order matrix groups is concerned, there is no difficulty in extending the parametrization to systems with n (> 2) inputs and outputs. For example, when n = 3, it can be shown that (e.g. see [MUR, Chapter 2]) any $U \in U(3)$ can be written as

$$U = R_{13}(\phi_1,\delta_1)R_{12}(\phi_2,\delta_2)R_{23}(\phi_3,\delta_3)\mathrm{diag}(e^{j\theta_1},e^{j\theta_2},e^{j\theta_3}) \qquad (2.5.1)$$

where

$$R_{13}(\phi_1,\delta_1) = \begin{bmatrix} c_1 & 0 & -\bar{s}_1 \\ 0 & 1 & 0 \\ s_1 & 0 & c_1 \end{bmatrix}$$

$$R_{12}(\phi_2,\delta_2) = \begin{bmatrix} c_2 & -\bar{s}_2 & 0 \\ s_2 & c_2 & 0 \\ 0 & 0 & 1 \end{bmatrix}$$

$$R_{23}(\phi_3,\delta_3) = \begin{bmatrix} 1 & 0 & 0 \\ 0 & c_3 & -\bar{s}_3 \\ 0 & s_3 & c_3 \end{bmatrix}$$

$$c_i := \cos\phi_i \; , \qquad s_i := e^{j\delta_i}\sin\phi_i \qquad\qquad i = 1,2,3$$

Any $G \in GL(3,\mathbb{C})$ can thus be parametrized by first doing an SVD and then applying (2.5.1) to the two unitary matrices of singular vectors in precisely the same way as was done in the proof of Theorem 2.1.4.

If we decompose a $G(s) \in \mathbb{R}(s)^{3\times 3}$ in the way just described to give

$$G(s) = Y(s)\Gamma(s)U(s)^*$$

then there will be 6 angles associated with each of the frame matrices, $U(s)$ and $Y(s)$. $\Gamma(s)$ will contain another 6 parameters, making up a total of 18, as expected (see §2.2). Because of the relatively large set of frame angles, it is not clear how to interpret them. Certainly, we still have a set of PG loci. However, for a reason to be given in the next section, we shall not pursue this line of development further.

§2.6 A Drawback of the Parameter Group Decomposition

We have given examples to show that the PG loci carry sensible information about the gain and phase behaviour of a system operator G(s). Unfortunately, it is also possible to construct examples to show that the PGD can be misleading as far as the input-output properties of a system are concerned.

Example 2.6.1

Suppose at some $s_o \varepsilon D_{NYQ}$, $G(s_o)$ has a PGD given by

$$G(s_o) = \begin{bmatrix} 0.1 & -0.995e^{-j\pi/2} \\ 0.995e^{j\pi/2} & 0.1 \end{bmatrix} \begin{bmatrix} 2 & 0 \\ 0 & 1 \end{bmatrix} \begin{bmatrix} 0.1 & -0.995 \\ 0.995 & 0.1 \end{bmatrix}^* \qquad (2.6.1)$$

We may represent $G(s_o)$ diagrammatically as follows:

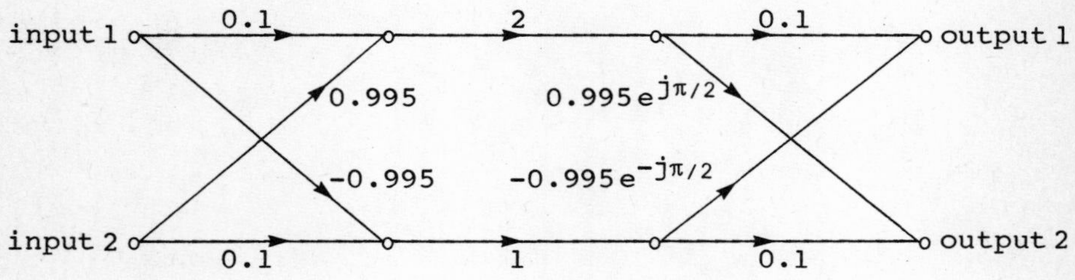

Fig.2.4

The PGD (2.6.1) associates zero phase to the PG loci for transmission from input 1 to output 1. However, because of the presence of large off-diagonal terms in both the input and output frame matrices, the transmission from input 1 to output 1 is dominated by the path with gain $(-0.995)(1)(-0.995e^{-j\pi/2})$ and it is apparent that at this particular s_o, the phase information is contained in the frame angles rather than the PG loci.

Having to take the frame angles into account when interpreting the behaviour of the PG loci is unsatisfactory. In the next chapter we therefore investigate another type of locus which has the great advantage that one can use a single measure which replaces the role of the frame angles.

CHAPTER 3 ALIGNMENT, NORMALITY AND QUASI-NYQUIST LOCI

§3.1 Frame Alignment and Normality

Some elementary properties of normal matrices and the reason why they are important in terms of spectral insensitivity have been given in §1.6. The characteristic value and Schur triangular decompositions are used in the treatment given there. In order to further relate normality to the singular value and polar decompositions, we now introduce the concept of frame alignment (in particular, alignment between the input and output singular-vector frame matrices, and the phase frame matrices).

Definition 3.1.1

(1) Two unitary (frame) matrices $Y, U \in \mathbb{C}^{m\times m}$ are said to be aligned if $\exists\ \Theta = \mathrm{diag}(e^{j\theta_1},..,e^{j\theta_m})$ s.t.

$$U^*Y = \Theta \tag{3.1.1}$$

(2) Let $G \in \mathbb{C}^{m\times m}$. If all possible SVD's of G:

$$G = Y\Sigma U^*$$

are such that Y, U are aligned, then G is said to be aligned. Otherwise G is misaligned.

Regarding the columns of Y and U as basis vectors of the spaces they span, the directions defined by the orthonormal bases of two aligned matrices are the same (since $Y = U\Theta$) except for phase factors between corresponding columns. Alignment is a rather strong condition and, as shown below, implies normality.

Prop 3.1.2

(1) G is aligned $\Rightarrow$ G is normal.

(2) If G has distinct singular values, then the converse of (1) holds.

Proof:

(1) In the notation of Definition 3.1.1, if G is aligned, then

$$\begin{aligned} G &= Y\Sigma U^* \\ &= Y\Sigma (U^*Y)Y^* \\ &= Y\Sigma\Theta Y^* \\ &= Y\Lambda Y^* \qquad \text{where } \Lambda := \Sigma\Theta \end{aligned}$$

Hence, by Theorem 1.6.1, G is normal.

(2) If G is normal, then it is unitarily similar to its diagonal matrix of eigenvalues,

$$\begin{aligned} G &= W\Lambda W^* \\ &= W|\Lambda|(\arg\Lambda)W^* \\ &= W|\Lambda|U^* \qquad \text{where } U := W(\arg\Lambda^*) \end{aligned}$$

which may be taken as an SVD of G. Clearly W, U are aligned. Moreover, if G has distinct singular values, then the fact that the singular-vector frames are aligned is independent of the particular SVD taken (see Prop 1.3.2(2)) and hence G is aligned. □

A consequence of the proposition is that the set of aligned matrices is a subset of the set of normal matrices. That the inclusion is proper can be demonstrated by an example.

Let $G = \begin{bmatrix} 0 & k \\ k & 0 \end{bmatrix}$ for some real $k > 0$, then G is normal because $GG^* = G^*G$. But G has an (non-unique) SVD given by

$$G = \begin{bmatrix} 1 & 0 \\ 0 & 1 \end{bmatrix} \begin{bmatrix} k & 0 \\ 0 & k \end{bmatrix} \begin{bmatrix} 0 & 1 \\ 1 & 0 \end{bmatrix}^*$$

and clearly G is not aligned.

For convenience in the subsequent discussion a matrix G having some equal singular values will be described as <u>partially gain isotropic</u>. (If all the singular values were equal it would be said to be <u>gain isotropic</u>. However it is convenient to take the term partially gain isotropic as including the case where possibly all singular values are the same.) The reason for the terminology is that any vector u lying in the subspace spanned by those input directions corresponding to t(> 1) equal singular values (say $\sigma_i = \cdots = \sigma_{i+t-1} = \sigma$) will satisfy $\|Gu\| = \sigma \|u\|$. Prop 3.1.2(2) can then be stated equivalently as: a normal matrix which is not aligned is partially gain isotropic. All the above ideas can be expressed in terms of an appropriate Venn diagram (see Fig.3.1)

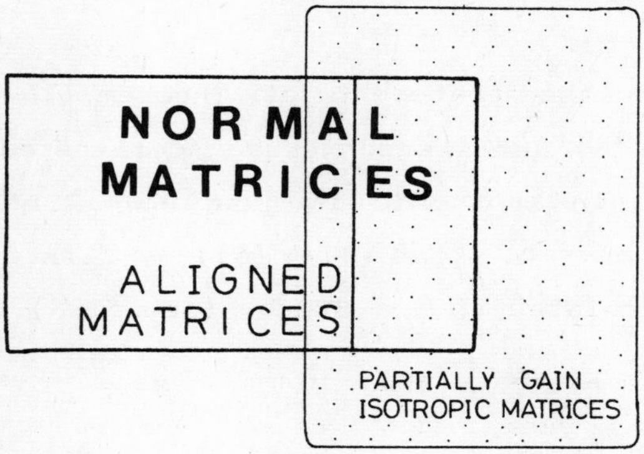

Fig.3.1 Venn diagram showing:

$\{\text{Aligned matrices}\} \subset \{\text{Normal matrices}\}$

$\{\text{Normal matrices}\} - \{\text{Aligned matrices}\} \subset \left\{\begin{array}{l}\text{Partially gain}\\ \text{isotropic matrices}\end{array}\right\}$

We shall now summarize the relationships between the various possible characterizations of normality and alignment. Let $G \in \mathbb{C}^{m\times m}$ and consider the following decompositions:

$$G = W\Lambda W^{-1} \qquad \text{(CVD)} \qquad (3.1.2a)$$

$$= S(D+T)S^* \qquad \text{(STD)} \qquad (3.1.2b)$$

$$= Y\Sigma U^* \qquad \text{(SVD)} \qquad (3.1.2c)$$

$$\left.\begin{array}{l} = \Phi M_r = P\Theta P^* U\Sigma U^* \qquad (3.1.2d) \\ = M_\ell \Phi = Y\Sigma Y^* P\Theta P^* \qquad (3.1.2e) \end{array}\right\} \text{(PD)}$$

Corresponding to each decomposition, there is a condition for G to be normal, or aligned.

Theorem 3.1.3

Let $G \in \mathbb{C}^{m\times m}$. In the notation of (3.1.2a-e) we have

(1) (Def) G is normal, i.e. $GG^* = G^*G$

$\Leftrightarrow$ (2) (CVD) G has an orthonormal eigenframe, i.e. W is unitary

$\Leftrightarrow$ (3) (STD) In (3.1.2b), $T = 0$

$\Leftarrow$ (4) (SVD) G is aligned

$\Leftrightarrow$ (5) (PD) Φ commutes with M_r, M_ℓ; and the phase frame matrix P of any PD can always be chosen so that P, Y, U are pairwise aligned

Moreover, if G is not partially gain isotropic, then (3) $\Rightarrow$ (4).

Proof:

(1) $\Leftrightarrow$ (2) is the statement of Theorem 1.6.1. (2) $\Leftrightarrow$ (3) is obvious. (4) $\Rightarrow$ (1) and its converse under the condition that G is not partially gain isotropic is just Prop 3.1.2. (5) $\Rightarrow$ (4) is trivial. It remains to prove (4) $\Rightarrow$ (5). We first note that the PD's (3.1.2d,e) are related to any SVD by (see §1.5)

$$\Phi = YU^* , \quad M_r = U\Sigma U^* , \quad M_\ell = Y\Sigma Y^*$$

If G is aligned, the phase matrix Φ can be decomposed as

$$\Phi = Y(U^*Y)Y^* = Y\Theta Y^* \tag{3.1.3}$$

$$= U(U^*Y)U^* = U\Theta U^* \tag{3.1.4}$$

for some diagonal unitary Θ. Using (3.1.4), then gives that

$$M_r\Phi = U\Sigma\Theta U^* = \Phi M_r$$

Similarly, using (3.1.3), we have that $M_\ell\Phi = \Phi M_\ell$. Again from (3.1.3) or (3.1.4), it is obvious that P can be chosen so that P, Y, U are pairwise aligned. □

Now normal or aligned matrices are only relatively small sets compared with "approximately normal" or "approximately aligned" matrices, which will be more important in our context. It is thus interesting to note that each of the five conditions of Theorem 3.1.3 suggests a way of measuring how close to normality or alignment a given matrix G is.

(1) By definition, a normal matrix commutes with its conjugate transpose. Thus if G is not normal, it is natural to measure the lack of commutativity in terms of the commutator $(GG^* - G^*G)$. We define (also see [WIL, Chapter 3 §50])

$$\Delta(G) := \frac{\|GG^* - G^*G\|^{1/2}}{\|G\|} \qquad (3.1.5)$$

(2) In terms of the CVD, non-normality is associated with non-orthogonality of the eigenframe W, for which one has the condition number

$$\kappa(W) := \|W\|_2 \|W^{-1}\|_2 \qquad (3.1.6)$$

(3) For the STD, we have already defined a measure of skewness (see §1.6) by

$$MS(G) = \frac{\|T\|}{\|G\|} \qquad (3.1.7)$$

(4) If $G = Y\Sigma U^*$ is misaligned, then U^*Y is not diagonal and it is reasonable to measure the misalignment by the departure of U^*Y from a nearest diagonal matrix. Specifically, we define the singular-vector-frame misalignment (or simply frame misalignment) of G to be

$$m(G) := \sup_{Y,U} \min_{\theta_i} \|U^*Y - \mathrm{diag}(e^{j\theta_i})\|_2 \qquad (3.1.8)$$

where the supremum is taken over all possible SVD's of G.

(5) For the PD, we can measure the lack of commutativity of Φ with M_r, M_ℓ by the norms of the appropriate commutators:

$$\|\Phi M_r - M_r\Phi\|/\|G\| \text{ and } \|\Phi M_\ell - M_\ell\Phi\|/\|G\|$$

Among the quantities defined in (1), (2) and (3), we shall stick to MS(G) as a measure of lack of normality because MS(G) is readily related to the spectral insensitivity of the matrix (see §1.6). As to (4) and (5), we shall use m(G) to quantify misalignment, for reasons to be given in the next few sections. In passing, we note that $\Delta(G)$

and MS(G) are related by an inequality (see [HEN, Theorem 1] or [WIL, Chapter 3 §50]):

$$MS(G) \leqslant \left(\frac{m^3 - m}{12} \right)^{1/4} \Delta(G) \qquad (3.1.9)$$

§3.2 Relationship between Skewness and Misalignment

To be consistent with the measures we have chosen for skewness and misalignment in the last section, whenever we say that a matrix G is approximately normal (resp. approximately aligned) it is to be interpreted as: G has a small skewness measure MS(G) (resp. a small misalignment measure m(G)). In this sense, Prop 3.1.2(1) can be extended to

" Approximate alignment $\Rightarrow$ Approximate normality "

To be more precise, we have:

Prop 3.2.1

Given any $\delta > 0$, $\exists$ $\varepsilon(\delta) > 0$ s.t.

$$m(G) < \delta \Rightarrow MS(G) < \varepsilon(\delta) \qquad (3.2.1)$$

and

$$\lim_{\delta \to 0} \varepsilon(\delta) = 0 \qquad (3.2.2)$$

□

It will be more convenient to prove this proposition after some terms are introduced in the next section. The proof is given in Appendix B.

§3.3 The Quasi-Nyquist Decomposition (QND)

Let $G \in \mathbb{C}^{m \times m}$ have an SVD

$$G = Y \Sigma U^* \qquad (3.3.1)$$

If G has distinct singular values, then the frame misalignment is independent of the particular SVD taken and (3.1.8) becomes

$$m(G) := \min_{\theta_i} \| U^*Y - \mathrm{diag}(e^{j\theta_i}) \|_2 \qquad (3.3.2)$$

Now suppose $\| U^*Y - \mathrm{diag}(e^{j\theta_i}) \|_2$ attains its minimum at some $\Theta = \mathrm{diag}(e^{j\theta_1}, .., e^{j\theta_m})$, then since Θ is unitary, we have

$$m(G) = \| U^*Y\Theta^* - I \|_2 \qquad (3.3.3)$$

Since

$$m(G) \leq \| U^*Y\Theta^* \|_2 + \| I \|_2$$

then clearly

$$0 \leq m(G) \leq 2 \qquad (3.3.4)$$

We now make an important observation: the diagonal matrix Θ obtained from the minimization problem (3.3.2) enables us to assign phases to the singular values. First define

$$\Gamma = \mathrm{diag}(\gamma_1, .., \gamma_m) := \Theta\Sigma \qquad (3.3.5)$$

$$Z := Y\Theta^* \qquad (3.3.6)$$

Then (3.3.1) can be written

$$G = (Y\Theta^*)(\Theta\Sigma)U^*$$

$$= Z\Gamma U^* \qquad (3.3.7)$$

Note that Θ is not related to the principal phases except in the case that G is normal. It is convenient however to use the symbol Θ in this more general sense in what follows. We shall call (3.3.7) the <u>quasi-Nyquist decomposition (QND)</u> of G. Note that in a QND, the diagonal matrix Γ is in general complex and the frame matrices Z, U are just the singular - vector frame matrices rescaled columnwise by some phase factors.

Substituting (3.3.6) into (3.3.3), we have that for the QND (3.3.7),

$$m(G) = \| U^*Z - I \|_2 \qquad (3.3.8)$$

It remains to solve (3.3.2) for a minimizing Θ. In the particular case that $V := U^*Y$ is 2×2, an explicit solution can be obtained. No such explicit solution has been found for the general case where Θ must be determined by a minimization procedure. We have for the 2×2 case:

Prop 3.3.1

Let $V \in U(2)$ be written as (see Prop 2.1.3)

$$V = \begin{bmatrix} \cos\phi & -e^{-j\delta}\sin\phi \\ e^{j\delta}\sin\phi & \cos\phi \end{bmatrix} \begin{bmatrix} e^{j\theta_1} & 0 \\ 0 & e^{j\theta_2} \end{bmatrix} \quad \text{where} \quad -\pi/2 \leq \phi \leq \pi/2$$

then $\min_{D \text{ diag, unitary}} \| V - D \|_2$ is solved by $\Theta = \text{diag}(e^{j\theta_1}, e^{j\theta_2})$.

Moreover $\| V - \Theta \|_2 = 2\left|\sin\left(\frac{\phi}{2}\right)\right|$ □

The proof is straightforward but tedious and is given in Appendix B. It follows from this proposition that for $G \in \mathbb{C}^{2\times 2}$

$$0 \leq m(G) \leq 2\sin\left(\frac{\pi}{4}\right) = \sqrt{2}$$

§3.4 Eigenvalue Bounds and the QND

The quasi-Nyquist decomposition of a matrix $G \in \mathbb{C}^{m\times m}$, together with the frame misalignment $m(G)$, can be used to localize the eigenvalues of G within regional bounds.

Let G have a QND

$$G = Z\Gamma U^* \tag{3.4.1}$$

Consider $\lambda \in \lambda(G)$. Then $(\lambda I - G)$ is singular, so that

$$\begin{aligned} 0 &= \det(\lambda I - G) \\ &= \det(\lambda I - Z\Gamma U^*) \\ &= \det Z \cdot \det(\lambda Z^* U - \Gamma) \cdot \det U^* \end{aligned}$$

Assuming that $(\lambda I - \Gamma)$ is nonsingular, we have

$$0 = \det[\lambda(\lambda I - \Gamma)^{-1}(Z^*U - I) + I]$$

which implies

$$\begin{aligned} 1 &\leq \|\lambda(\lambda I - \Gamma)^{-1}(Z^*U - I)\|_2 \\ &\leq \|\lambda(\lambda I - \Gamma)^{-1}\|_2 \|Z^*U - I\|_2 \end{aligned} \tag{3.4.2}$$

Now

$$\begin{aligned} \|Z^*U - I\|_2 &= \|U^*Z - I\|_2 \\ &= m(G) \qquad \text{by (3.3.8)} \end{aligned}$$

Since $\lambda(\lambda I - \Gamma)^{-1}$ is a diagonal matrix, its spectral norm is simply given by the diagonal entry having the maximum modulus, say, the ith diagonal element. (3.4.2) can then be written

$$1 \leq \left|\frac{\lambda}{\lambda - \gamma_i}\right| \cdot m(G) \tag{3.4.3}$$

Using the fact that $|\lambda| \leq \gamma_{max}$, where $\gamma_{max} := \sigma_{max}(G) = \|G\|_2$ is the maximum singular value of G, a disk bound for λ follows immediately,

$$|\lambda - \gamma_i| \leq |\lambda| \cdot m(G) \leq \gamma_{max} \cdot m(G) \tag{3.4.4}$$

Remember that we have assumed $(\lambda I - \Gamma)$ to be nonsingular. If this is not true, then $\lambda = \gamma_i$ for some i and (3.4.4) is also satisfied. Hence (3.4.4) always holds.

Since λ may be any eigenvalue of G, we have shown that each eigenvalue must lie in a disk $D(\gamma_i ; \gamma_{max} \cdot m(G))$ centred at some γ_i with radius $\gamma_{max} \cdot m(G)$.

(3.4.4) usually will not give a good bound, particularly when $|\lambda| << \gamma_{max}$. In order to get a tighter bound, we start from (3.4.3) again.

Let $\lambda = x + jy$ and $\gamma_i = a + jb$, then

$$\begin{aligned} & (3.4.3) \\ \Leftrightarrow\ & |\lambda - \gamma_i|^2 \leq |\lambda|^2 m(G)^2 \\ \Leftrightarrow\ & (x-a)^2 + (y-b)^2 \leq (x^2 + y^2) m(G)^2 \\ \Leftrightarrow\ & (1 - m(G)^2) x^2 - 2ax + (1 - m(G)^2) y^2 - 2by + a^2 + b^2 \leq 0 \end{aligned} \tag{3.4.5}$$

Case 1 :

If $m(G) = 1$, then (3.4.5) degenerates into a linear inequality,

$$2(ax + by) \geq a^2 + b^2$$

$$\Leftrightarrow \frac{a}{\sqrt{a^2+b^2}} x + \frac{b}{\sqrt{a^2+b^2}} y \geq \frac{1}{2}\sqrt{a^2+b^2}$$

$$\Leftrightarrow \cos\angle\gamma_i \, x + \sin\angle\gamma_i \, y \geq \frac{1}{2}|\gamma_i| \tag{3.4.6}$$

(3.4.6) restricts $\lambda = x + jy$ to a half plane, as shown in Fig.3.2(a).

Case 2 :

If $m(G) < 1$, then (3.4.5) can be written

$$x^2 - \frac{2a}{1 - m(G)^2} x + y^2 - \frac{2b}{1 - m(G)^2} y \leq - \frac{a^2 + b^2}{1 - m(G)^2} \tag{3.4.7}$$

Completing squares for the left hand side, we have

$$\left(x - \frac{a}{1 - m(G)^2}\right)^2 + \left(y - \frac{b}{1 - m(G)^2}\right)^2 \leq \frac{(a^2 + b^2)\, m(G)^2}{(1 - m(G)^2)^2}$$

$$\Leftrightarrow \left|(x + jy) - \frac{a + jb}{1 - m(G)^2}\right|^2 \leq \frac{|a + jb|^2\, m(G)^2}{(1 - m(G)^2)^2}$$

$$\Leftrightarrow \left|\lambda - \frac{\gamma_i}{1 - m(G)^2}\right| \leq \frac{|\gamma_i|\, m(G)}{|1 - m(G)^2|} \tag{3.4.8}$$

(3.4.8) bounds λ inside a disk (see Fig.3.1(b)) :

$$\lambda \in D\left(\frac{\gamma_i}{1 - m(G)^2} ; \frac{|\gamma_i|\, m(G)}{|1 - m(G)^2|} \right) \tag{3.4.9}$$

Case 3 :

If $m(G) > 1$, then all the steps for case 2 carry over except that the inequality signs in (3.4.7 - 8) should be reversed. Accordingly, we finally deduce that λ lies outside an open disk (see Fig.3.2(c)) :

$$\lambda \notin D\left(\frac{\gamma_i}{1 - m(G)^2} ; \frac{|\gamma_i|\, m(G)}{|1 - m(G)^2|} \right)^\circ \tag{3.4.10}$$

where $D(\cdot,\cdot)^\circ$ denotes the interior of the disk $D(\cdot,\cdot)$.

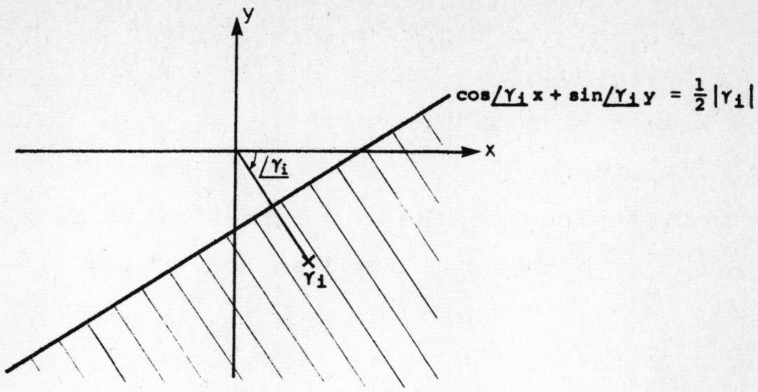

(a) Case 1: If m(G) = 1, $\lambda = x + jy$ satisfying (3.4.6) lies in the shaded region.

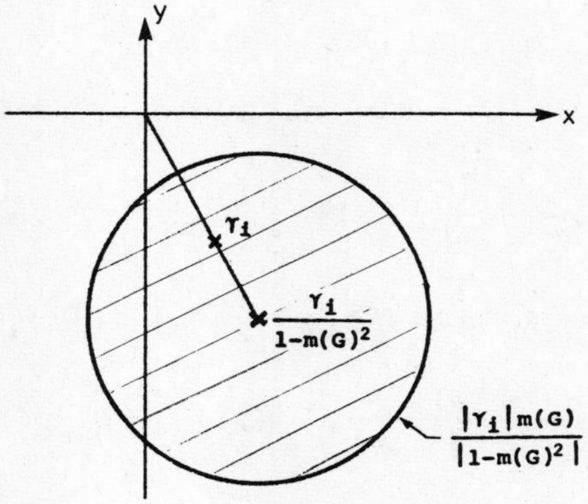

(b) Case 2: If m(G) < 1, λ lies inside the disk (3.4.9).

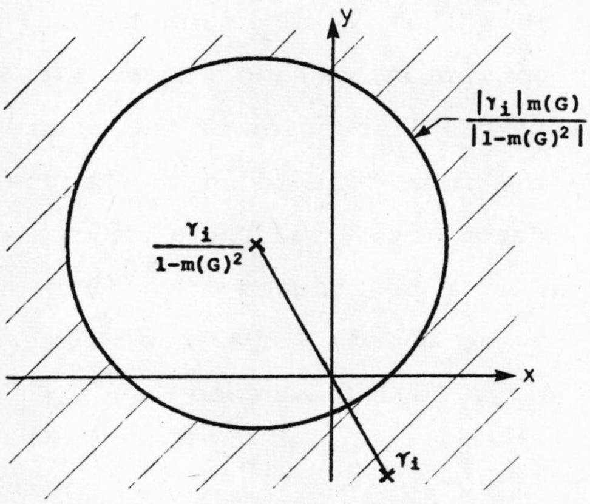

(c) Case 3: If m(G) > 1, λ lies outside the open disk (3.4.10).

Fig.3.2 Regional bounds for an eigenvalue λ of G.

Note that in both cases 2 and 3, the centre of the disk lies on the line through the origin and γ_i. But in case 2, the centre is on the same side of the origin as γ_i while in case 3, it is on the opposite side.

Combining one of the three possibilities (3.4.6), (3.4.9) or (3.4.10) with the previous disk bound (3.4.4) we have

Prop 3.4.1

Let $G = Z\Gamma U^* \in \mathbb{C}^{m\times m}$ be any decomposition of G satisfying $Z, U \in U(m)$ and $\Gamma = \mathrm{diag}(\gamma_1, \ldots, \gamma_m)$.

Also let $\delta = \|U^*Z - I\|_2$

and $\gamma_{max} = \sigma_{max}(G) = \|G\|_2$

then for any $\lambda \in \lambda(G)$, $\exists$ a diagonal element γ_i of Γ s.t. $\lambda \in B_i$ where

$$B_i := \begin{cases} D\left(\dfrac{\gamma_i}{1-\delta^2}\,;\,\dfrac{|\gamma_i|\,\delta}{|1-\delta^2|}\right) \cap D(\gamma_i\,;\,\gamma_{max}\,\delta) & \text{if } 0 \leq \delta < 1 \\ \{x+jy \mid \cos\angle\gamma_i\, x + \sin\angle\gamma_i\, y \geq \tfrac{1}{2}|\gamma_i|\} \cap D(\gamma_i\,;\,\gamma_{max}) & \text{if } \delta = 1 \\ \left[\mathbb{C} - D\left(\dfrac{\gamma_i}{1-\delta^2}\,;\,\dfrac{|\gamma_i|\,\delta}{|1-\delta^2|}\right)^\circ\right] \cap D(\gamma_i\,;\,\gamma_{max}\,\delta) & \text{if } 1 < \delta \leq 2 \end{cases}$$

□

It is easy to show that $\gamma_i \in B_i$ and that if $\gamma_i \neq 0$, then $0 \notin B_i$. Note that in the proposition, we have not restricted the decomposition of G to a QND because the discussion of the bounds holds for any other decomposition of the form defined in the proposition. However, if the QND is used, then $\delta = m(G)$ will be a minimum which means that the regional bounds B_i will be reduced to a minimum area.

In addition to the bounds of Prop 3.4.1, we also have that $\gamma_{max} \geq |\lambda| \geq \gamma_{min} (:= \sigma_{min}(G))$ and hence

$$\lambda \in D(0\,;\,\gamma_{max}) \cap [\mathbb{C} - D(0\,;\,\gamma_{min})^\circ]$$

This annular inclusion region, however, is not related to the quantity δ, as the B_i's are.

§3.5 Quasi-Nyquist Loci

Since $G(s) \in \mathbb{R}_p(s)^{m\times m}$ has no poles on the Nyquist D-contour D_{NYQ}, we can apply the QND to G(s) for each $s \in D_{NYQ}$, and get

$$G(s) = Z(s)\Gamma(s)U(s)^* \tag{3.5.1}$$

where $$\Gamma(s) = \mathrm{diag}\big(\gamma_1(s),\ldots,\gamma_m(s)\big) \tag{3.5.2}$$

and Z(s), U(s) $\in$ U(m) satisfy

$$m(G(s)) = \| U(s)^* Z(s) - I \|_2 \tag{3.5.3}$$

As s traverses D_{NYQ}, $\gamma_i(s)$ (i = 1,...,m) trace out a set of m loci. The combined loci will be referred to as the quasi-Nyquist loci (QN loci or simply QNL). The loci of $\gamma_i(s)$ will be labelled by QNLi in diagrams. Now for each point on $\gamma_i(s)$, we can define a region $B_i(s)$ by Prop 3.4.1 with δ set to m(G(s)). Thus each $\gamma_i(s)$ carries a band of area, along its locus, swept out by $B_i(s)$. Moreover, it is clear, by Prop 3.4.1, that the characteristic loci must lie in the union of the bands of area defined by $B_i(s)$ $(i = 1,\ldots,m; s \in D_{NYQ})$.

Example 3.5.1 (QN loci of Gas Turbine: AUTM)

The QN loci of AUTM (see Example 2.3.1 and Appendix C) are given in Fig.3.3(a,b). The bands of area swept out by $B_i(j\omega)$ (i = 1,2), as shown in Fig.3.3(c,d), clearly enclose the characteristic gain loci shown in Fig.3.3(e). At each frequency s = jω, the size of the region $B_i(j\omega)$ is determined by m(G(jω)), which is given in Fig.3.3(f). Note that for the system AUTM, m(G(jω)) remains reasonably small (< 0.5) for all $s \in D_{NYQ}$. Also note that the $B_i(j\omega)$ diminishes towards high frequency. This is because $\gamma_{max} \to 0$ as $s \to \infty$ for strictly proper systems.

Example 3.5.2 (QN loci of Chemical Reactor: REAC)

The system considered is a 2-input, 4-state, 2-output chemical reactor model which will be referred to as REAC. This example was carefully studied in [MAC2] and a listing of details of REAC is given in Appendix D. The appropriate set of diagrams for REAC is given in Fig.3.4(a - f). Unlike the previous example, REAC has a rather larger frame misalignment, particularly at high frequencies.

The regional bounds put around the Quasi-Nyquist loci can be interpreted as an indicator of the degree of misalignment of a system over its effective bandwidth. If $G(s_o)$ is aligned, then the regions $B_i(s_o)$ reduce to points and the Quasi-Nyquist loci coincide with the characteristic gain loci, at $s = s_o$. If $G(s_o)$ is approximately aligned having a small $m(G(s_o))$, then the regions $B_i(s_o)$ should be small and the Quasi-Nyquist loci will be close to the characteristic gain loci. Evidently it is desirable for the bands around the QN loci to be narrow. This is also a sufficient, though not necessary, checking condition for a system to be near normality over its effective bandwidth (see Prop 3.2.1).

In general, given a system $G(s)$, there is no reason to expect that the frames of $G(s)$ are approximately aligned. As is typical of most other bounds for eigenvalues, the regional bound defined by the $B_i(s)$ is likely to be conservative. Because of this the QN loci are not intended to replace the characteristic gain loci as an analysis tool for general systems.

As far as spectral insensitivity is concerned, it clearly would be useful to incorporate into a design technique ways of approximately normalizing the system as part of the design objective. We shall discuss how this can be done, at a single frequency (in §3.6) and over a frequency band (in Chapters 4 through 6). Although normality is not quite the same as alignment, the chances are that normalizing a system often aligns its frames (for most normal matrices are aligned; see Prop 3.1.2(2)). The QN loci, taken with bands of $B_i(s)$, will then serve as a sufficient check for judging the results achieved.

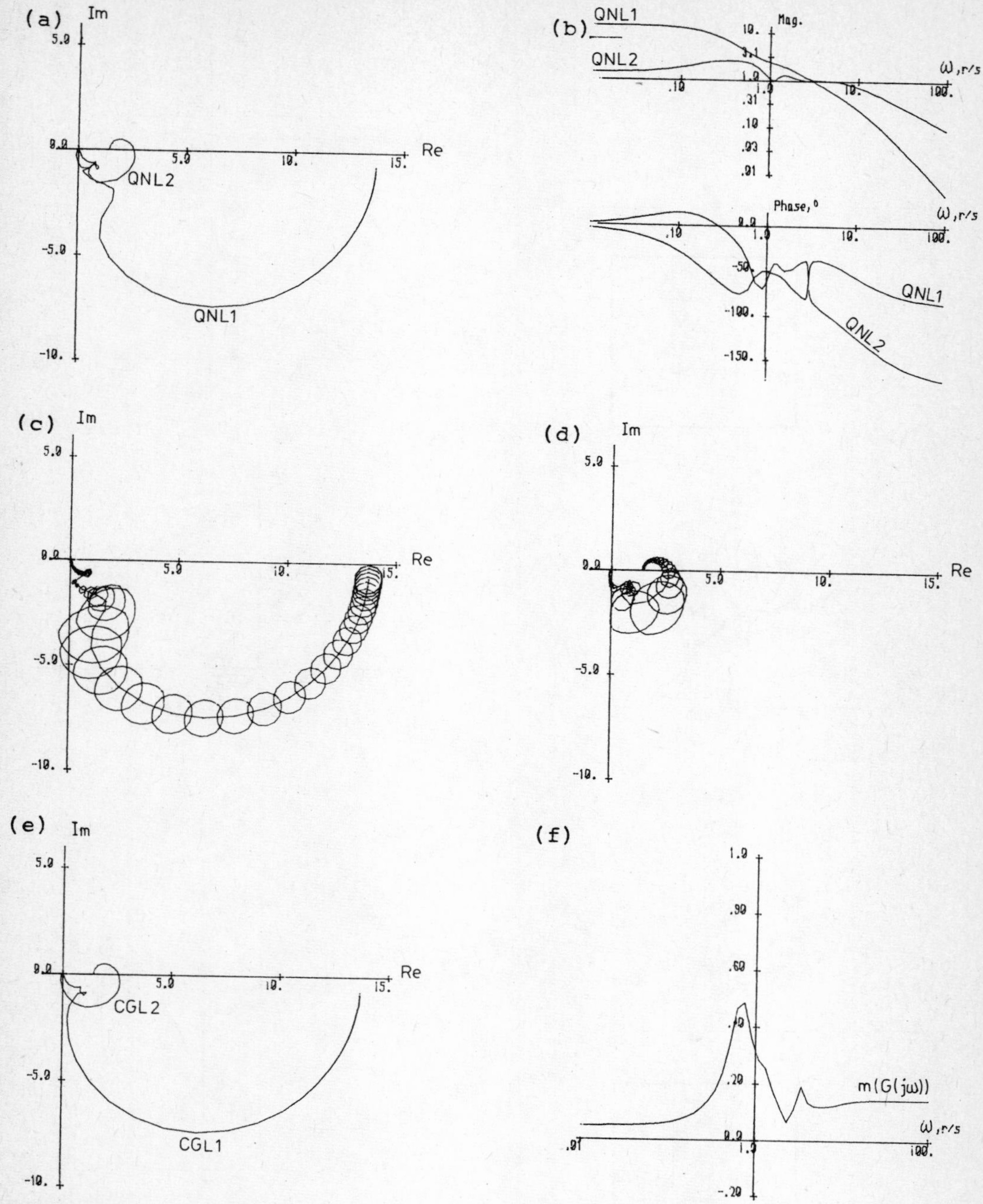

Fig.3.3

(a),(b) QN loci of the system AUTM.

(c),(d) QNLi taken with the bands $B_i(j\omega)$ (i=1,2).

(e) Characteristic gain loci.

(f) Frame misalignment $m(G(j\omega))$.

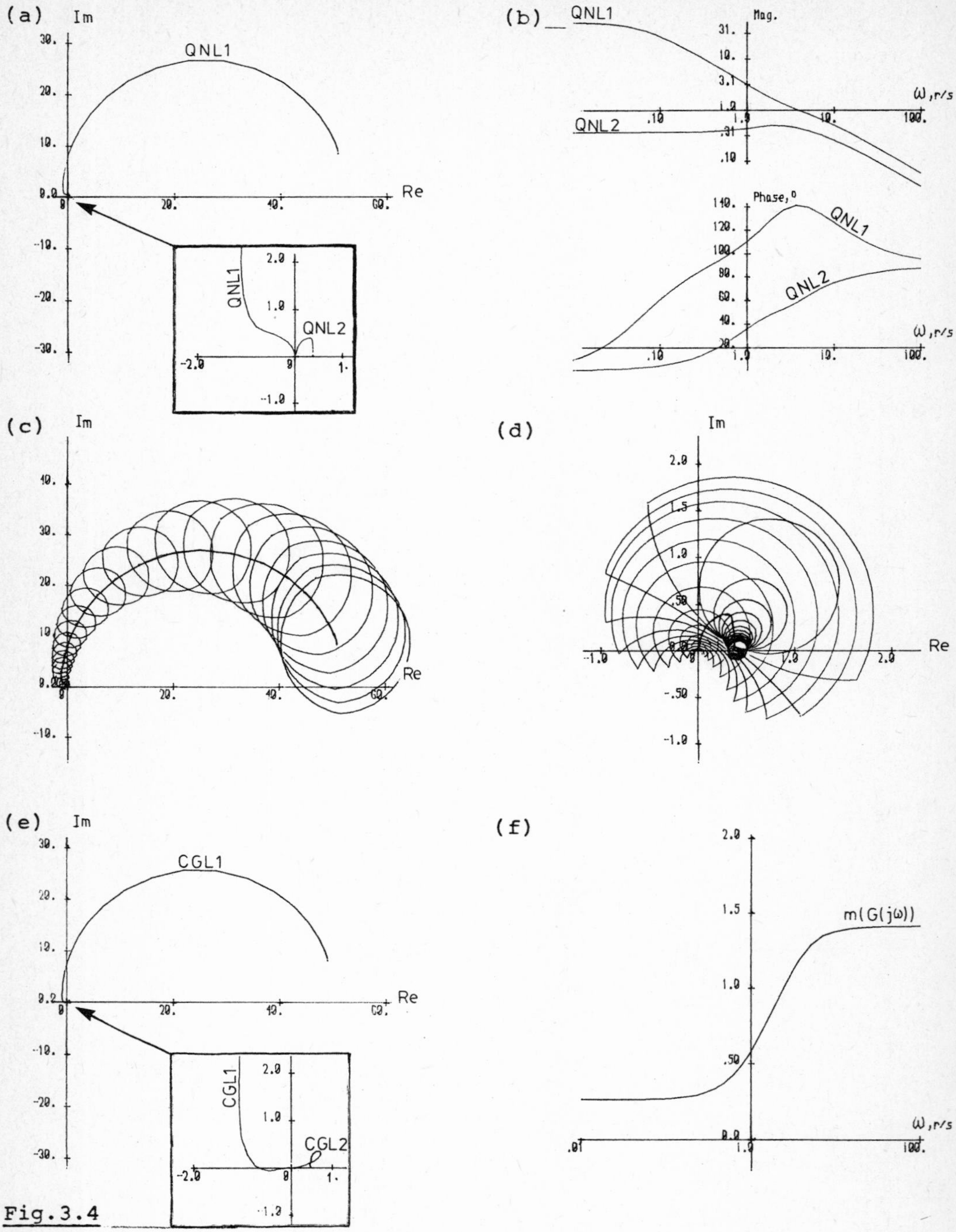

Fig.3.4

(a),(b) QN loci of the system REAC.

(c),(d) QNLi taken with the bands $B_i(j\omega)$ $(i=1,2)$.

(e) Characteristic gain loci.

(f) Frame misalignment $m(G(j\omega))$.

§3.6 Standardization at $s = 0$ or ∞

As an initial step in a design procedure, or else for the purpose of gaining insight into a system's behaviour, it is useful to have ways of normalizing a system at some specially chosen frequencies.

Lack of normality, or skewness, is associated with a "reversed-frame" operator in the sense that if at some $s_\alpha \in \mathbb{C}$, $G(s_\alpha)$ has an SVD given by

$$G(s_\alpha) = Y_\alpha \Sigma_\alpha U_\alpha^* \tag{3.6.1}$$

then we can annihilate the skewness by multiplying $G(s_\alpha)$ with the matrix $U_\alpha \mathrm{diag}(\cdot) Y_\alpha^*$ to get

$$G(s_\alpha)\left[U_\alpha \mathrm{diag}(\cdot) Y_\alpha^*\right] = Y_\alpha \left[\Sigma_\alpha \mathrm{diag}(\cdot)\right] Y_\alpha^* \tag{3.6.2}$$

This leads one to the idea of standardizing a system $G(s)$ by orthogonalizing its eigenframe, at some particular frequency, using a constant precompensator.

§3.6.1 Standardization at $s = 0$

Since at $s = 0$, $G(0)$ is real, it has a real SVD

$$G(0) = Y_o \Sigma_o U_o^T \tag{3.6.3}$$

in which the frames Y_o, U_o are orthogonal. Hence we can define a (real) constant precompensator by

$$K_o(k_1, \ldots, k_m) := U_o \,\mathrm{diag}(k_i)\, Y_o^T \tag{3.6.4}$$

for some set of $k_i \in \mathbb{R}$. In particular, if each $k_i = 1$ or -1, then K_o $(:= K_o(k_1, \ldots, k_m))$ will be an orthogonal matrix. We say that $G(s)K_o$ is <u>a standardization of $G(s)$ at $s = 0$</u> if $k_i = \pm 1 \;\forall i$, or <u>a standardization with gain balancing at $s = 0$</u> if $k_i \neq \pm 1$ for some i. This precompensated system has the following properties:

(1) Since

$$G(0)K_o = Y_o\left(\Sigma_o \,\mathrm{diag}(k_i)\right) Y_o^T$$

$G(s)K_o$ has an orthogonal eigenframe at $s=0$.

(2) If K_o is orthogonal, then the principal gains (singular values) of $G(s)$ and $G(s)K_o$ are the same.

(3) The diagonal elements of $\Sigma_o \operatorname{diag}(k_i)$ are the eigenvalues of $G(0)K_o$. Since $\Sigma_o \operatorname{diag}(k_i)$ is real, the characteristic gain loci of $G(s)K_o$ start from the real axis. Moreover, the ith characteristic gain locus will start from the positive or negative real axis according as $k_i > 0$ or < 0. A perhaps oversimplified approach for choosing the signs of k_i is that if $G(s)$ has no poles or zeros in $\mathbb{C}_+$, then we require all the characteristic loci to start from the positive side of the real axis. However, if $G(s)$ has open-loop unstable poles, then some characteristic loci may have to start from the negative real axis to produce the required anticlockwise encirclements for closed-loop stability.

§3.6.2 Standardization at $s=\infty$

An alternative frequency at which it is feasible to orthogonalize the eigenframe by precompensating with a real matrix is $s=\infty$. Asymptotically, as $|s|\to\infty$, $G(s)$ takes the form [HUN]

$$G(s) \simeq Y(s)\operatorname{diag}\left(\frac{\gamma_i}{s^{r_i}}\right) U(s)^* \tag{3.6.5}$$

where: the γ_i's are real,

the r_i's are orders of infinite zeros of $G(s)$,

and $Y(s)$, $U(s) \to$ orthogonal matrices Y_∞, U_∞ respectively.

Hence if we define

$$K_\infty(k_1,\ldots,k_m) := U_\infty \operatorname{diag}(k_i)\, Y_\infty^T \tag{3.6.6}$$

for some $k_i \in \mathbb{R}$, then K_∞ ($:= K_\infty(k_1,\ldots,k_m)$) is real. We say that $G(s)K_\infty$ is <u>a standardization (with gain balancing) at $s=\infty$</u> if $k_i = \pm 1\ \forall i$ (correspondingly $k_i \neq \pm 1$ for some i). $G(s)K_\infty$ has the following properties:

(1) Since

$$G(j\omega)K_\infty \rightarrow Y_\infty \mathrm{diag}\left(\frac{\gamma_i k_i}{(j\omega)^{r_i}}\right) Y_\infty^T \qquad \text{as } s = j\omega \rightarrow j\infty$$

the precompensated system has an orthogonal eigenframe at $s = \infty$.

(2) If each $k_i = 1$ or -1, then K_∞ will be orthogonal, leaving the principal gains of $G(s)K_\infty$ the same as those of $G(s)$.

(3) If the infinite zeros are of the same order, i.e. $r_1 = r_2 = \ldots = r_m$, then by appropriate choice of the gains k_i, it is possible to "balance up" the characteristic gains in a neighbourhood of $s = \infty$ (see Example 3.6.1 below).

(4) Since the characteristic gains behave asymptotically as $(\gamma_i k_i/(j\omega)^{r_i})$, the phases of the characteristic gain loci will approach $\pm r_i(\pi/2)$.

We now give two examples of standardization at $s = 0$ or ∞.

Example 3.6.1 (A system with no poles or zeros in $\mathbb{C}_+$)

Consider

$$G(s) = 5\begin{bmatrix} \dfrac{1}{s+1} & \dfrac{1}{s+4} \\ \dfrac{2}{(s+1)(s+3)} & \dfrac{-3s}{(s+2)(s+4)} \end{bmatrix}$$

which has open-loop poles at $\{-1,-2,-3,-4\}$, finite zeros at $\{-0.409,-3.26\}$ and two first-order infinite zeros.

The CGL, QN loci and frame misalignment of G(s) are given in Fig.3.5. As G(s) has no poles in $\mathbb{C}_+$, the characteristic gain loci should not encircle $(-1 + j0)$ for closed-loop stability (see Theorem 1.2.1), and it is apparent that CGL2 starts from the "wrong side" of the real axis.

Following (3.6.3) and (3.6.4), a low frequency standardization is performed as follows. First do an SVD of G(0),

$$G(0) = \begin{bmatrix} 5 & 1.25 \\ 3.33 & 0 \end{bmatrix} = \underbrace{\begin{bmatrix} .843 & .538 \\ .538 & -.843 \end{bmatrix}}_{Y_o} \begin{bmatrix} 6.10 & 0 \\ 0 & .683 \end{bmatrix} \underbrace{\begin{bmatrix} .985 & -.173 \\ .173 & .985 \end{bmatrix}^T}_{U_o^T}$$

So one obtains:

$$K_o(1,1) = U_o Y_o^T = \begin{bmatrix} .737 & .676 \\ .676 & -.737 \end{bmatrix}$$

Graphs for $G(s)K_o$ are given in Fig.3.6. Note that K_o has brought CGL2 to start from the positive real axis and that $m(G(s)K_o)$ is greatly reduced over the low frequency band.

Next, we use (3.6.5) and (3.6.6) to do a high frequency precompensation with gain balancing, as follows. As $s = j\omega \to j\infty$,

$$G(j\omega) \to \underbrace{\begin{bmatrix} -.347 & -.938 \\ .938 & -.347 \end{bmatrix}}_{Y_\infty} \begin{bmatrix} \frac{15.9}{j\omega} & 0 \\ 0 & \frac{4.72}{j\omega} \end{bmatrix} \underbrace{\begin{bmatrix} -.109 & -.994 \\ -.994 & .109 \end{bmatrix}^T}_{U_\infty^T}$$

and we put

$$K_\infty(k_1,k_2) = U_\infty \operatorname{diag}(k_1,k_2)\, Y_\infty^T = \begin{bmatrix} 1.73 & .577 \\ 0 & -.577 \end{bmatrix} \qquad (3.6.7)$$

where $k_1 = \frac{k}{15.9}$ $\qquad k_2 = \frac{k}{4.72}$ $\qquad k = \sqrt{15.9\times 4.72}$

Graphs for $G(s)K_\infty$ are given in Fig.3.7. Note that K_∞ has balanced up the gains in a neighbourhood of the infinite frequency and that $m(G(s)K_\infty)$ is small over the high frequency band.

The present system, having no poles or zeros in $\mathbb{C}_+$ and being quite well-behaved from the analysis we have done so far, is expected to be easy to control. For example, appropriately using K_o over low frequencies and K_∞ over high frequencies, a possible simple PI (proportional plus integral) controller is

$$K_{PI}(s) = K_\infty + \frac{1}{s} K_o(1,3)$$

where K_∞ is as given in (3.6.7) and instead of using the orthogonal $K_o(1,1)$, we have injected an extra gain factor of 3 in the direction of small gain. Graphs of $G(s)K_{PI}(s)$ are given in Fig.3.8; these correspond to an acceptable closed-loop performance.

Example 3.6.2 (A system with a zero in $\mathbb{C}_+$)

Consider

$$G(s) = 5 \begin{bmatrix} \frac{1}{s+1} & \frac{1}{s+4} \\ \frac{2}{(s+1)(s+3)} & \frac{s}{(s+2)(s+4)} \end{bmatrix}$$

which has open-loop poles at {-1,-2,-3,-4}, finite zeros at {-2.56,1.56} and two first-order infinite zeros.

The CGL, QN loci and frame misalignment of G(s) are given in Fig.3.9. Standardizing at $s=0$ by $K_0(1,1)$ gives the results shown in Fig.3.10. Standardizing at $s=\infty$ by $K_\infty(1,-1)$ gives the results shown in Fig.3.11. Note that because G(s) has a zero in $\mathbb{C}_+$, one of the characteristic gain loci has an extra 180° phase lag. By specifying the diagonal gains k_1, k_2 of $K_\infty(k_1,k_2)$ to be 1 and -1, we have specifically required the two characteristic gain loci of $G(s)K_\infty(1,-1)$ to approach the origin along the negative and the positive imaginary axis (corresponding to 90° and 270° phase lag). Comparison of Fig.3.9(e), Fig.3.10(e) and Fig.3.11(e) shows that for this example, standardizing at either end of the frequency scale tends to upset frame alignment at the other frequency end. Not surprisingly, a simple PI-controller of the form $K_\infty(\cdot,\cdot) + \frac{1}{s}K_0(\cdot,\cdot)$ does not work well for this system.

§3.7 Diagonalizing at a Critical Frequency

We can use a real constant precompensator to remove skewness either at $s_\alpha = 0$ or ∞ because at these frequencies, $G(s_\alpha)$ has a decomposition of the form (3.6.1) in which the input and output frames U_α and Y_α are real. At any other intermediate frequency s_α, the matrices U_α and Y_α are usually complex, and so it is generally not possible to construct a real precompensator of the form $U_\alpha \operatorname{diag}(k_i) Y_\alpha^*$. Yet as an initial design step, it is often more important to be able to deal with a critical frequency near the cross-over region than to standardize at $s=0$ or ∞. Moreover, in order to reduce closed-loop interaction without using excessive gains near the cross-over region (see §4.3.2), it is necessary to diagonalize (or approximately diagonalize) the system rather than just to orthogonalize the eigenframe. As this cannot be done exactly by a constant precompensator, we have recourse to optimization of a suitable approximation. Specifically, at a chosen critical frequency s_α, a precompensator K_α can be constructed in such a way that $G(s_\alpha)K_\alpha$ is as close as possible, in a least-squares sense, to a diagonal phase matrix $D_p = \mathrm{diag}(e^{j\theta_i})$. That is, K_α is a solution to the minimization problem:

$$\min_{K_\alpha, \theta_i} \| G(s_\alpha)K_\alpha - \mathrm{diag}(e^{j\theta_i}) \|^2 \qquad (3.7.1)$$

Direct techniques of calculus show that (3.7.1) has an explicit solution given by (see [EDM3, §4], [MAC3, §4.1]):

$$\begin{cases} K_\alpha = [\mathrm{Re}(G^*G)]^{-1}[\mathrm{Re}(G^*\mathrm{diag}(e^{j\theta_i}))] \\ \theta_i = \frac{1}{2} \arg x_{ii} \end{cases}$$

where we have denote $G(s_\alpha)$ simply by G and x_{ii} are the diagonal elements of

$$X := G[\mathrm{Re}(G^*G)]^{-1}(G^*G)^T[\mathrm{Re}(G^*G)]^{-1}G^T$$

For example, such a preliminary precompensation (at $s_\alpha = 3j$) has already been built into the system AUTM, which is clear if one examines Fig.2.1 or Fig.3.3(b).

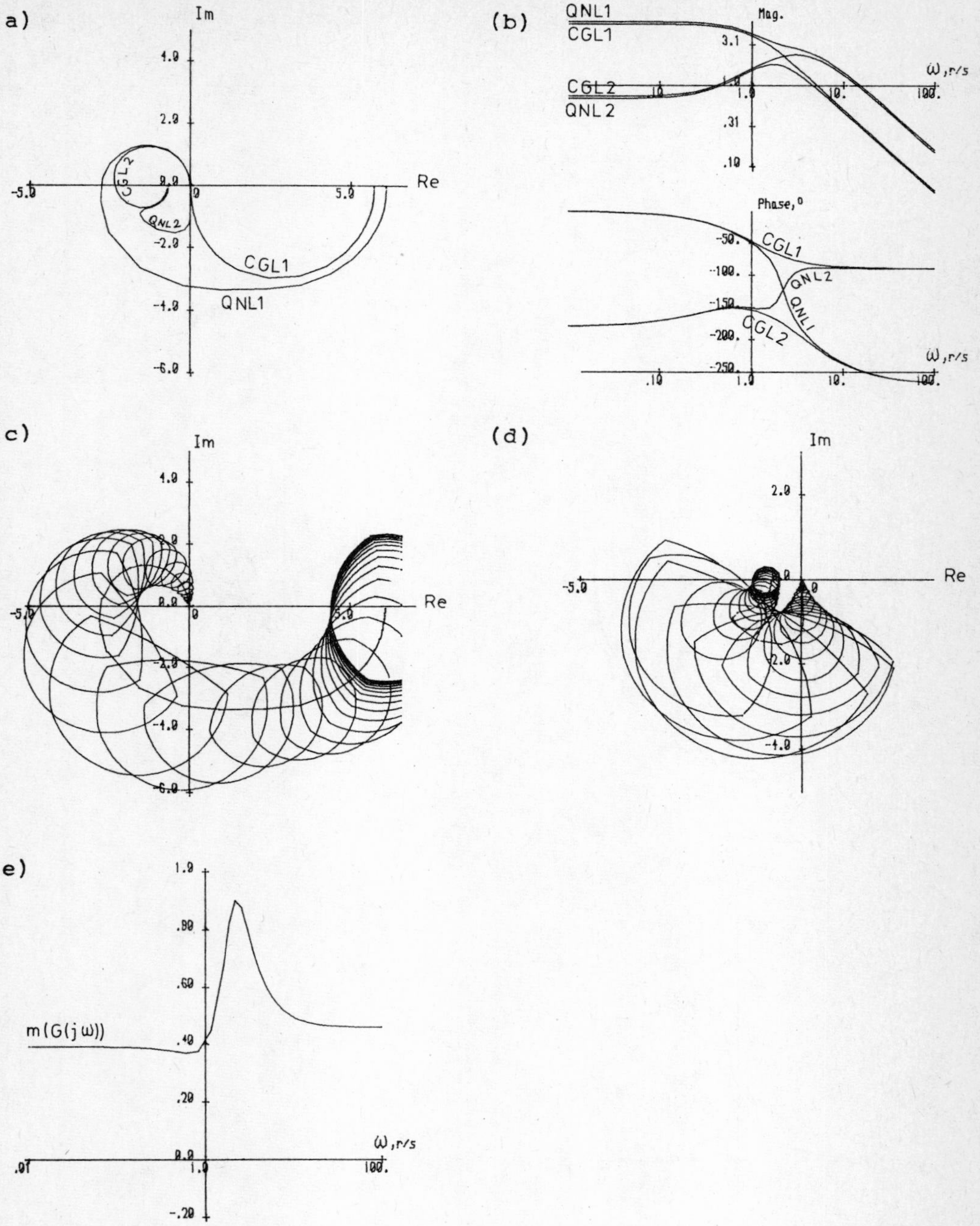

Fig.3.5 The uncompensated system G(s) of Example 3.6.1.

(a),(b) Characteristic gain loci and QN loci.

(c),(d) QNL1 and QNL2 with regional bounds.

(e) Frame misalignment.

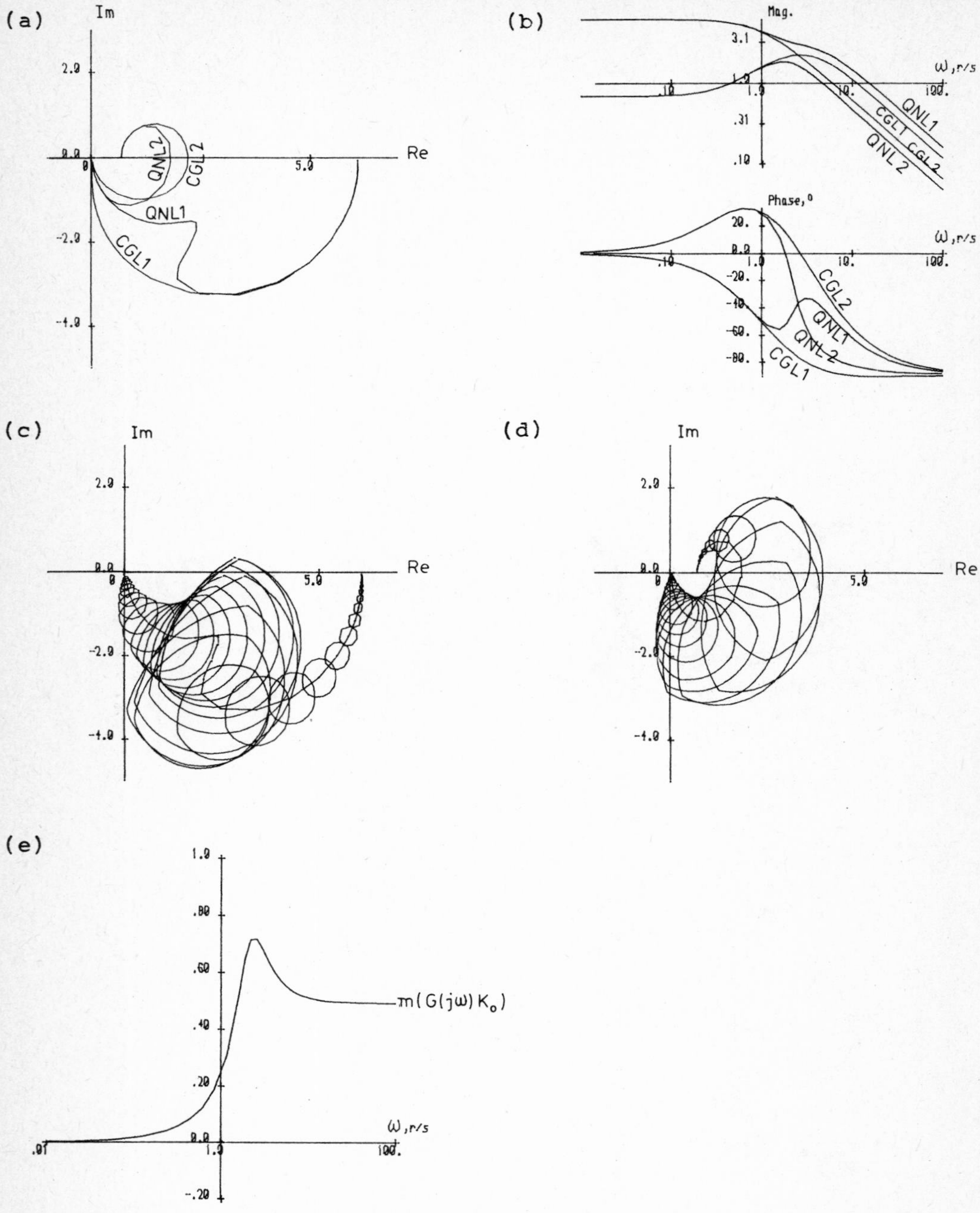

Fig.3.6 A standardization, $G(s)K_o(1,1)$, at $s=0$.

(a),(b) Characteristic gain loci and QN loci.

(c),(d) QNL1 and QNL2 with regional bounds.

(e) Frame misalignment.

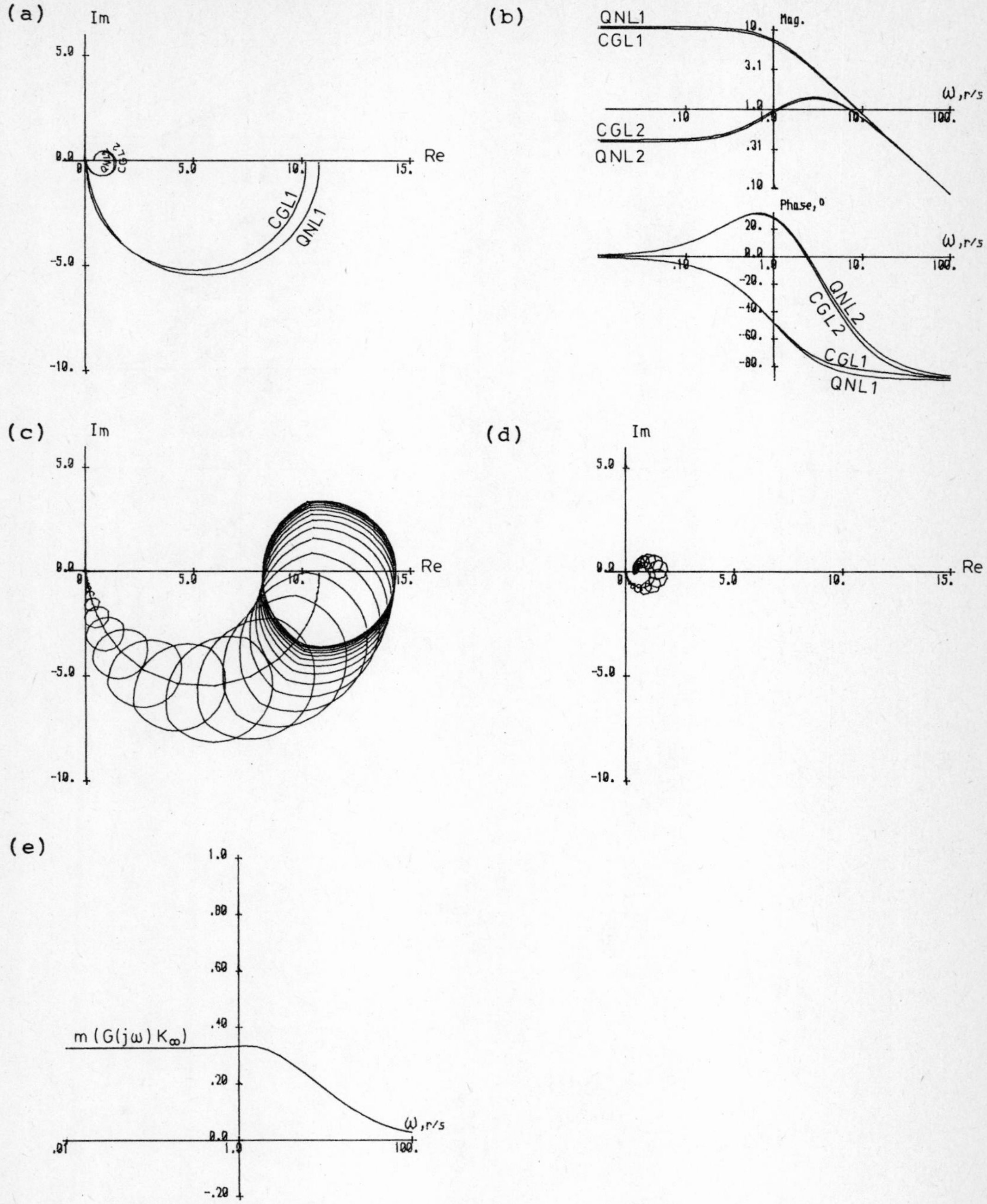

<u>Fig.3.7</u> A standardization, $G(s)K_{\infty}(k_1,k_2)$, with gain balancing at $s=\infty$.

(a),(b) Characteristic gain loci and QN loci.

(c),(d) QNL1 and QNL2 with regional bounds.

(e) Frame misalignment.

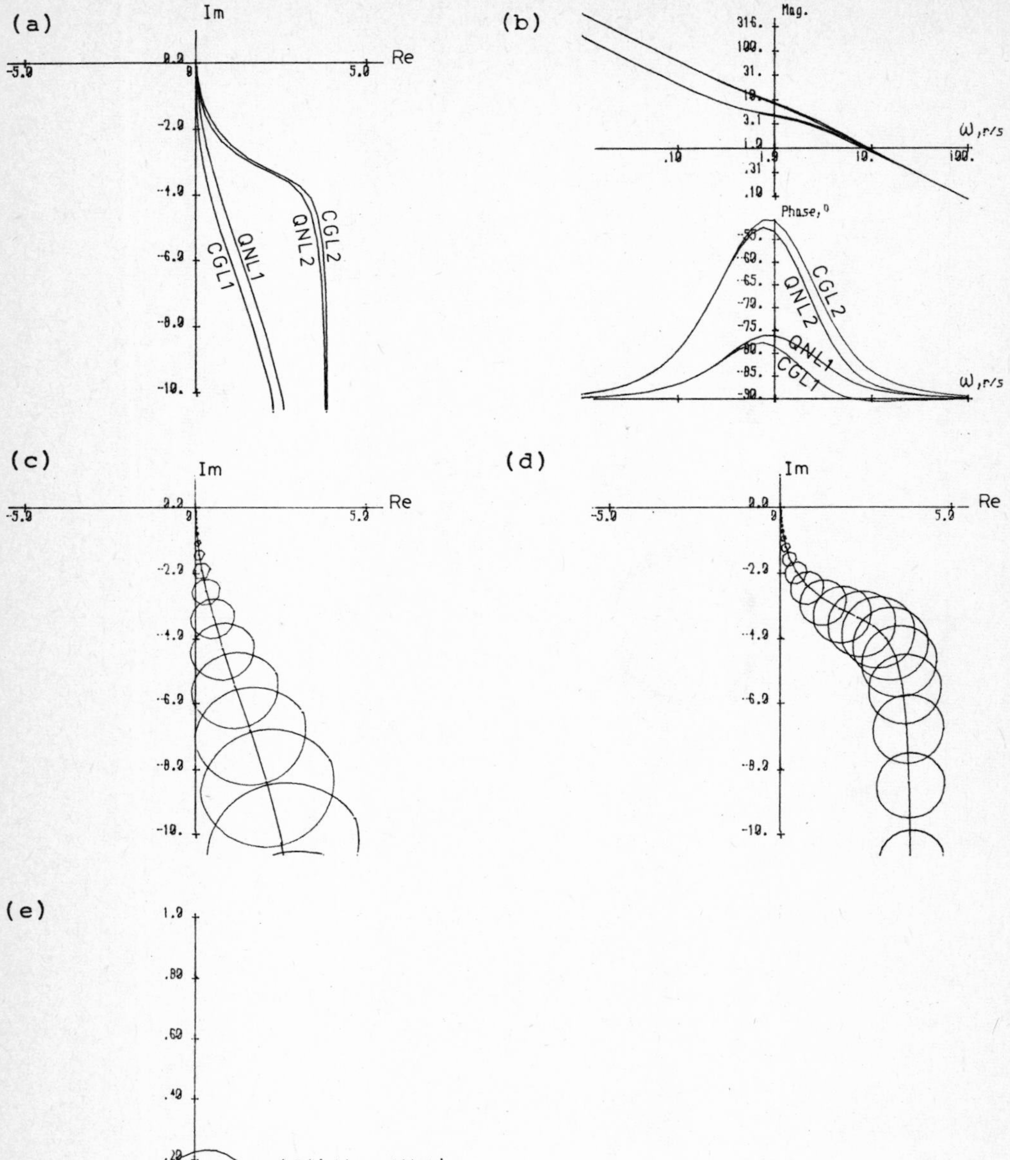

<u>Fig.3.8</u> The system precompensated by a PI-controller of the form $K_{PI}(s) = K_\infty(\cdot,\cdot) + \frac{1}{s}K_0(\cdot,\cdot)$.

(a),(b) Characteristic gain loci and QN loci.

(c),(d) QNL1 and QNL2 with regional bounds.

(e) Frame misalignment.

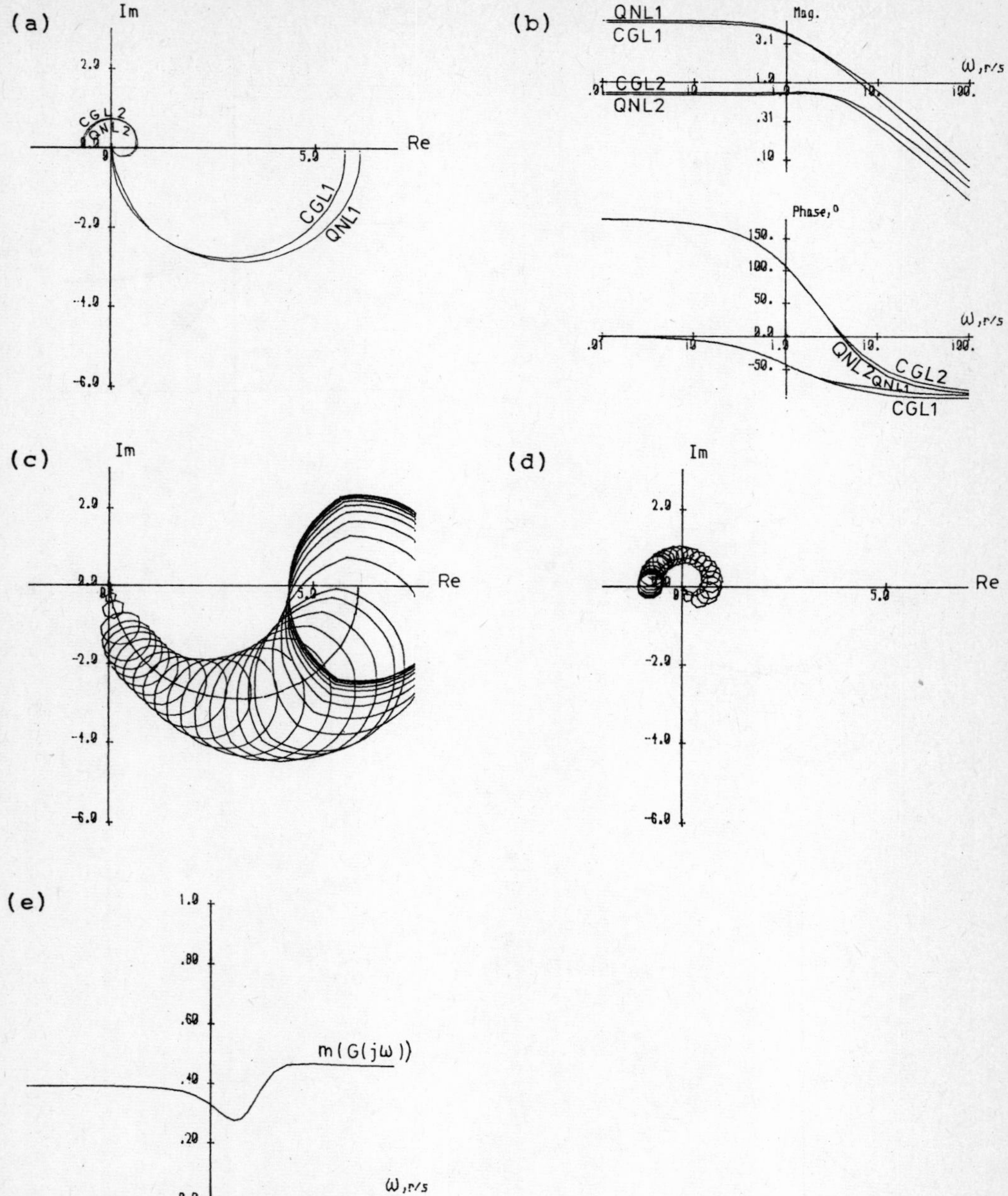

<u>Fig.3.9</u> The uncompensated system G(s) of Example 3.6.2.

(a),(b) Characteristic gain loci and QN loci.

(c),(d) QNL1 and QNL2 with regional bounds.

(e) Frame misalignment.

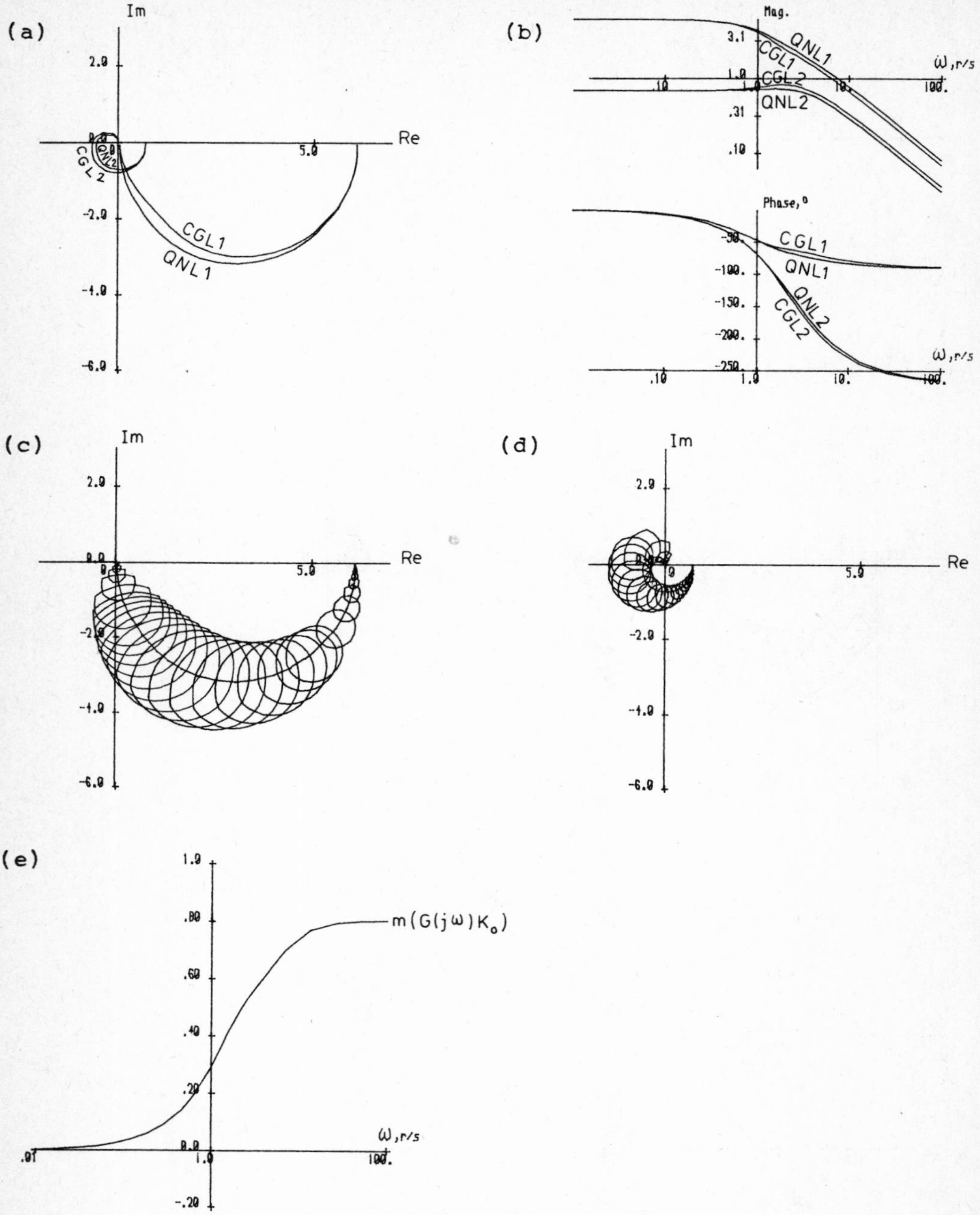

<u>Fig.3.10</u> A standardization, $G(s)K_o(1,1)$, at $s=0$.

(a),(b) Characteristic gain loci and QN loci.

(c),(d) QNL1 and QNL2 with regional bounds.

(e) Frame misalignment.

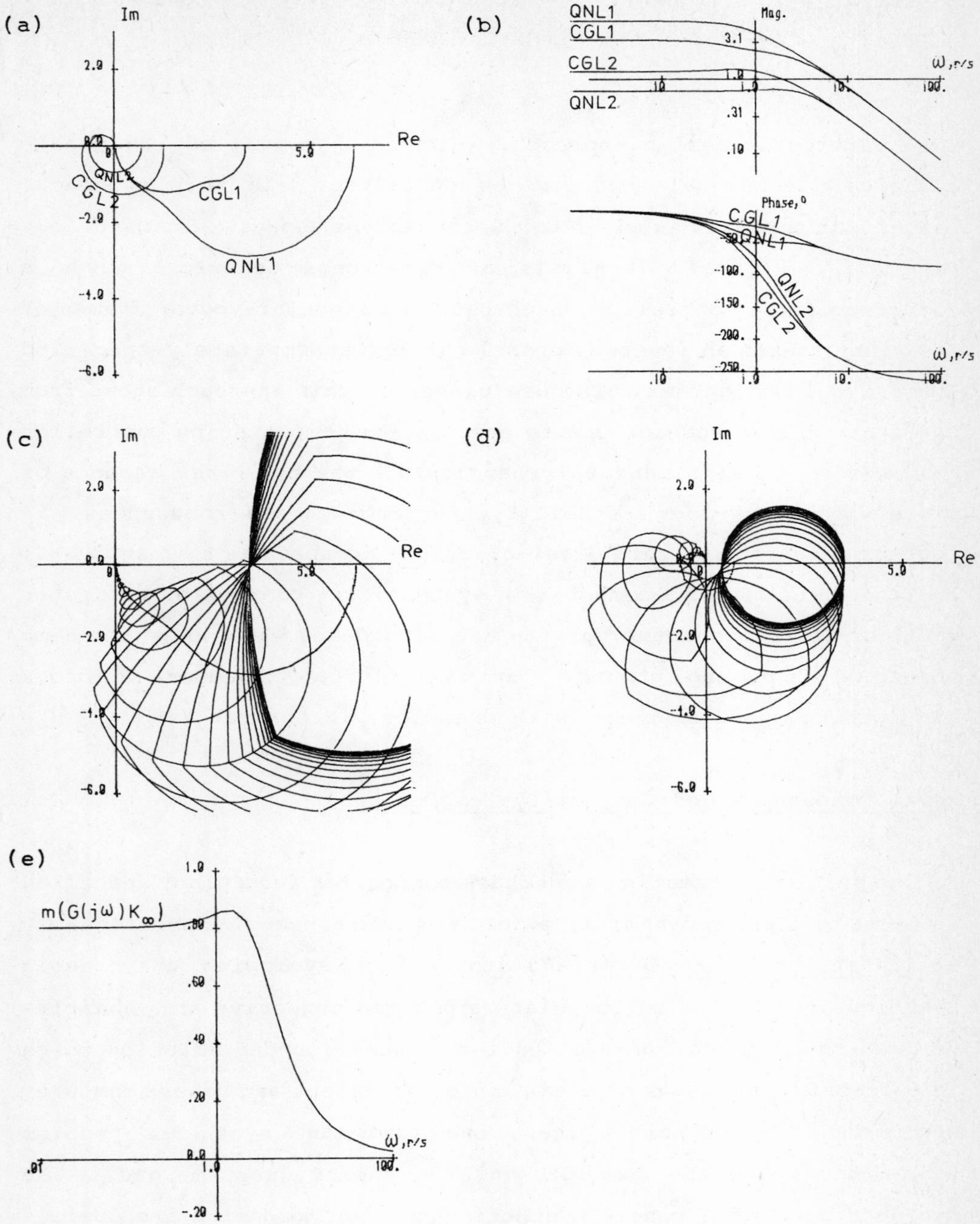

<u>Fig.3.11</u> A standardization, $G(s)K_\infty(1,-1)$, at $s=\infty$.

(a),(b) Characteristic gain loci and QN loci.

(c),(d) QNL1 and QNL2 with regional bounds.

(e) Frame misalignment.

CHAPTER 4 A QUASI-CLASSICAL DESIGN TECHNIQUE

A particular design approach, which we have called the Quasi-classical technique, will now be described. In this approach, particular emphasis is given to the robustness aspects of closed-loop behaviour. A careful analysis of robustness behaviour gives a structure for a controller which uses the singular-vector frames of the plant (taken in reversed order) but with appropriately-specified Quasi-Nyquist diagrams. The usefulness of this approach stems from the fact that it enables one to specify the compensating controller in a way which simultaneously handles all the three key aspects of closed-loop behaviour : stability, performance and robustness. A further advantage of this quasi-classical approach is that it is well suited to the use of a computer in synthesizing controller parameter values. In this chapter, plants are considered which have the same number of inputs and outputs. Plants with different numbers of inputs and outputs are discussed, with examples, in Chapter 7.

§4.1 Computer-Aided Control System Design

The problem of creating a feedback controller for a plant described in terms of a given dynamical model has three aspects, conventionally called analysis, synthesis and design. In developing a synthesis technique the aim is to formulate a desired objective as a sharply-defined mathematical problem having a well-founded solution which is expressible in terms of a workable, efficient and robust computer algorithm. In principle then, one loads the synthesis problem description into the computer and the answer duly emerges. The disadvantages of a purely synthetic approach to design are obvious in an engineering context since the role of the designer, particularly the exercise of his intuitive judgement and skill, is severely reduced. An even greater drawback is that, at the beginning of his investigations, the designer simply may be unable to specify what he wants in terms of a desired final system behaviour because he lacks vital information on what he will have to pay, in engineering terms, for the various aspects of his desired final system performance. In developing a design technique, one seeks to give a practising and experienced design engineer a set of manipulative and interpretative tools which will enable him to build up, modify and assess a design put together on the basis of physical reasoning within the guidelines laid down by his engineering experience. Thus design inevitably involves both analysis and synthesis and, in the development of design techniques, consideration of the way in which a designer interacts with a computer is vitally important. It is imperative to share the burden of work between computer and designer in such a way that each makes an appropriate contribution to the overall solution. In developing a computer-aided design technique the aims thus should be to:

(i) allow the designer to fully deploy his intuition, skill and experience while still making an effective use of powerful theoretical tools; and

(ii) to harness the manipulative power of the computer to minimize the level of detail with which the designer has to contend.

The designer communicates with the computer through an <u>interface</u>. This allows him to <u>interpret</u> what the computer has done and to <u>specify</u> what he wishes it to do next. In general terms we will call anything which is presented to the designer by the computer, and which is relevant to the design process, an <u>indicator</u>. The designer must operate within an appropriate <u>conceptual framework</u>, and any powerful interactive design package must present the designer with the full set of indicators required to specify his needs and interpret his results in the context of his conceptual framework.

The computer is used for calculation, manipulation and optimization. In any fully-developed interactive design package the "tuning" of the controller parameters is best done by systematic use of appropriate optimization techniques. Generally speaking, in the design process the designer will be doing analysis and the computer will be doing synthesis. That is to say the computer will be solving a series of changing and restrictively-specified synthesis problems put to it by the designer as he works his way through a range of alternatives, between which he chooses on the grounds of engineering judgement, as he travels towards his final design.

Since the designer will usually want to think in the most physical way possible about the complex issues facing him, a high premium is placed on developing a conceptual framework for him to work in which makes the maximum use of his spatial intuition, that is on one which is formulated as much as possible in geometric and topological terms. The particular approach developed here is based on generalizations of the gain and phase concepts and techniques of classical frequency-response-based feedback theory. The designer, after analyzing the open-loop plant characteristics, sets feasible specifications for a compensated plant which are approximately achieved by a computer synthesis procedure. In short the designer sets specifications and analyzes controllers, and the synthesis of controllers of prescribed structure is carried out by the computer.

§4.2 Stability

Closed-loop stability is assessed using the generalized Nyquist stability criterion : for a given loop-breaking point the corresponding closed-loop system will be stable if and only if the number of anti-clockwise encirclements of the critical point $(-1+j0)$ by the characteristic gain loci of the return-ratio transfer-function matrix for the break point is equal to the number of open-loop unstable poles of the return-ratio transfer-function matrix. If a scalar gain parameter k multiplies the loop-transmittance then the critical point becomes $(-\frac{1}{k}+j0)$.

Closed-loop stability is thus investigated by use of appropriate sets of characteristic gain loci (generalized Nyquist diagrams) or equivalent sets of Bode diagrams. Alternatively closed-loop stability could be investigated by using appropriate multivariable root locus diagrams, which contain exactly the same information as generalized Nyquist diagrams, packaged in a different form (frequency as a function of gain rather than gain as a function of frequency). Only generalized Nyquist and Bode diagrams will be used here.

§4.3 Performance

In considering the performance of a feedback control system one is concerned with :

(i) command tracking,

(ii) sensor noise rejection, and

(iii) disturbance rejection.

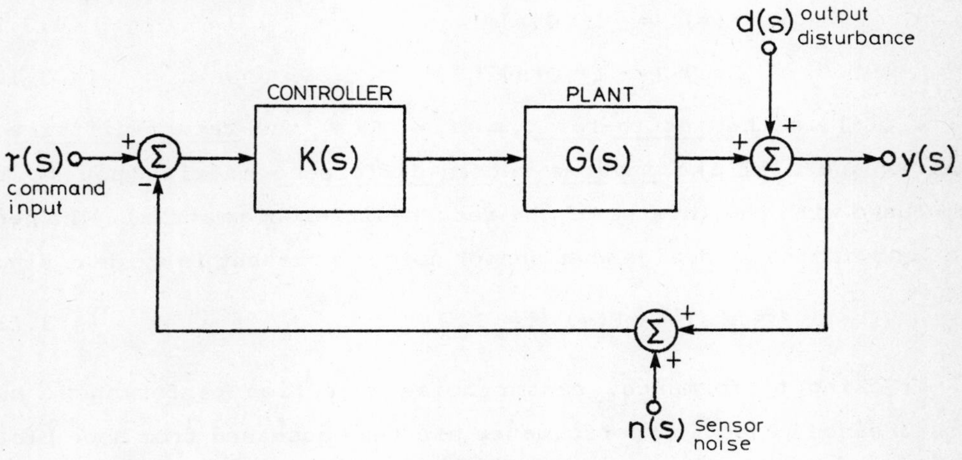

Fig.4.1

For the system of Fig.4.1, the plant output vector y(s) is the sum of responses to input commands r(s), sensor noise n(s) and output disturbances d(s). Calling these $y_r(s)$, $y_n(s)$ and $y_d(s)$ and computing them separately we have :

$$y_r(s) = [I + G(s)K(s)]^{-1}G(s)K(s)\,r(s) \qquad (4.3.1)$$

which, if G(s)K(s) is not identically singular, we may write in the form

$$y_r(s) = [I+(G(s)K(s))^{-1}]^{-1} r(s) \tag{4.3.2}$$

For the response to sensor noise we have

$$y_n(s) = -[I+G(s)K(s)]^{-1}G(s)K(s)\,n(s)$$
$$= -[I+(G(s)K(s))^{-1}]^{-1} n(s) \tag{4.3.3}$$

and the response to an output disturbance is given by

$$y_d(s) = [I+G(s)K(s)]^{-1} d(s) \tag{4.3.4}$$

These may be written in the form

$$y_r(s) = L(s)^{-1} r(s) \tag{4.3.5}$$
$$y_n(s) = -L(s)^{-1} n(s) \tag{4.3.6}$$
$$y_d(s) = F(s)^{-1} d(s) \tag{4.3.7}$$

where

$$Q(s) := G(s)K(s) \tag{4.3.8}$$
$$F(s) := I+Q(s) \tag{4.3.9}$$
$$L(s) := I+Q(s)^{-1} \tag{4.3.10}$$

Here Q(s) is the return-ratio matrix, F(s) the return-difference matrix and L(s) the inverse-return-difference matrix (not to be confused with the inverse of the return-difference matrix). The need for compromise in design when sensor noise is present is evident since

$$L(s)^{-1} + F(s)^{-1} = I \tag{4.3.11}$$

Tracking performance, sensor noise rejection performance, and disturbance rejection performance are thus assessed from Bode plots of the maximum and minimum principal gains for the relevant operators $L(s)^{-1}$ and $F(s)^{-1}$.

§4.3.1 Reversed-Frame-Normalizing (RFN) Controller

Let $G(s) \in \mathbb{R}_p(s)^{m\times m}$ have QN decomposition (see §3.3)

$$G(s) = Z(s)\Gamma_G(s)U(s)^* \tag{4.3.12}$$

If for some diagonal matrix $\Gamma_K(s)=\mathrm{diag}(\gamma_{K1}(s),\ldots,\gamma_{Km}(s))$, we define a precompensator to be

$$K(s) := U(s)\Gamma_K(s)Z(s)^* \quad \forall\ s \in D_{NYQ} \tag{4.3.13}$$

then the precompensated system is given by

$$\begin{aligned} Q(s) &:= G(s)K(s) \\ &= Z(s)\Gamma_G(s)U(s)^*U(s)\Gamma_K(s)Z(s)^* \\ &= Z(s)\Gamma_G(s)\Gamma_K(s)Z(s)^* \\ &= Z(s)\Gamma_Q(s)Z(s)^* \end{aligned} \tag{4.3.14}$$

where $$\begin{aligned} \Gamma_Q(s) &= \mathrm{diag}(\gamma_{Q1}(s),\ldots,\gamma_{Qm}(s)) \\ &:= \Gamma_G(s)\Gamma_K(s) \end{aligned} \tag{4.3.15}$$

A precompensator of the form given by equation (4.3.13) will be called a reversed-frame-normalizing (RFN) controller for the plant G(s). The name arises because:

(i) The singular-vector frames of K(s) are those of G(s) taken in reversed order; and

(ii) The resulting compensated plant Q(s) is normal.

The reason for wishing to achieve a controller of this form will become clearer as the detailed discussion of specifications unfolds. The essence of the matter is that, by seeking to synthesize a controller of this form, we can be sure of handling all three aspects of closed-loop behaviour — that is stability, performance and robustness — simultaneously and satisfactorily.

In general, it is of course not possible to realize K(s) exactly, as defined by (4.3.13), by a rational matrix. What we shall in fact do is find a rational matrix which reasonably approximates K(s). We shall discuss how to do this in considerable detail in the next two chapters. In this chapter, for the sake of simplicity of exposition, we assume that we can find a rational K(s) which approximates the right hand side of (4.3.13) to such a degree that for our purpose, we can neglect the difference between the non-rational $U(s)\Gamma_K(s)Z(s)^*$ and its rational approximation K(s).

From (4.3.14) and (4.3.15) we note that:

(1) $Q(s)$ is normal for $\forall s \in D_{NYQ}$.

(2) $\{\gamma_{Q1}(s),\ldots,\gamma_{Qm}(s)\}$ is the set of characteristic gain loci as well as the QN loci for $Q(s)$. In particular, the moduli of the eigenvalues of $Q(s)$ are equal to the principal gains (singular values). Hence the characteristic gain loci will give accurate information for both performance and stability.

We now proceed to discuss certain aspects of performance on the assumption that $K(s)$ has the particular form given in (4.3.13).

§4.3.2 Interaction

If the closed-loop transfer matrix (CLTM)

$$\begin{aligned} \text{CLTM} &:= \big(I + Q(s)\big)^{-1} Q(s) \\ &= Z(s)\big(I + \Gamma_Q(s)\big)^{-1}\Gamma_Q(s) Z(s)^* \\ &= Z(s)\,\text{diag}\Big(\frac{\gamma_{Qi}(s)}{1+\gamma_{Qi}(s)}\Big)_{i=1}^{m} Z(s)^* \end{aligned} \tag{4.3.16}$$

is not diagonal, then for some input transform vector

$$r(s) = (0,\ldots,0,r_i(s),0,\ldots,0)^T$$

having a single nonzero ith component, the output response is not necessarily restricted to the ith output. In many cases, a design objective is to make the ith output respond to the ith input alone and so to reduce the interaction between the ith input and $j(\neq i)$th output. To require low interaction is equivalent to saying that the off-diagonal terms of the CLTM should be small compared with the diagonal elements. For practical reasons, this has to be accomplished in different ways according to what frequency range is being considered. To be specific, we define the low, high and critical frequency ranges to be

$$\text{LFR} := \{\omega \in [0,\infty) \mid \sigma_{min}(Q(j\omega)) > 1\}$$

$$\text{HFR} := \{\omega \in [0,\infty) \mid \sigma_{max}(Q(j\omega)) < 1\}$$

$$\text{CFR} := \{\omega \in [0,\infty) \mid \sigma_{max}(Q(j\omega)) \geq 1 \geq \sigma_{min}(Q(j\omega))\}$$

First, consider $s = j\omega$ with $\omega \in \text{LFR}$. Write (4.3.16) as

$$\text{CLTM} = I - Z(j\omega)\,\text{diag}\Big(\frac{1}{1+\gamma_{Qi}(j\omega)}\Big)_{i=1}^{m} Z(j\omega)^* \qquad (4.3.17)$$

Two obvious bounds for off-diagonal terms are

$$\begin{bmatrix}\text{Sum of squares of moduli of} \\ \text{off-diagonal terms of CLTM}\end{bmatrix} \leq \sum_{i=1}^{m} \frac{1}{|1+\gamma_{Qi}(j\omega)|^2}$$

$$\begin{bmatrix}\text{Moduli of any off-diagonal} \\ \text{term of CLTM}\end{bmatrix} \leq \frac{1}{\sigma_{min}(Q(j\omega))-1}$$

Clearly, if $\sigma_{min}(Q(j\omega)) >> 1$, then both bounds will be small and the CLTM will be approximately diagonal. This justifies the widely accepted rule that feedback with uniformly high gains approximately decouples the closed-loop system. However, it is also well known that because of power considerations or stability reasons, high gains are not feasible at high frequencies.

Next consider $\omega \in \text{CFR}$, then some $\gamma_{Qi}(j\omega)$ has modulus less than one and the second term on the right of (4.3.17) will not be small. Returning to (4.3.16), we see that any one of the following conditions is sufficient for closed-loop non-interaction:

(i) $Z(j\omega)$ is diagonal. Recall that $Z(j\omega)$ is the output frame of $G(j\omega)$; thus this imposes a restriction on the system $G(s)$.

(ii) $\text{diag}\Big(\frac{\gamma_{Qi}(j\omega)}{1+\gamma_{Qi}(j\omega)}\Big)$ is a scalar matrix, i.e. $\gamma_{Qi}(j\omega)=\cdots=\gamma_{Qm}(j\omega)$.

These two conditions say that to reduce critical frequency interaction, we can either

(i) do a preliminary precompensation to make $G(j\omega)$ approximately diagonal (e.g. by the technique of §3.7) so that $G(j\omega)$ will have a diagonal output frame; or

(ii) "balance up" the $\gamma_{Qi}(j\omega)$'s of the compensated system.

Finally, if $\omega \varepsilon$ HFR and if $\sigma_{max}(Q(j\omega)) << 1$, then we shall not attempt to reduce interaction on the grounds of the negligible transmission through the system and the possibly large uncertainty of the model at such frequencies.

§4.3.3 Tracking Accuracy and Disturbance Rejection

The ability of the outputs to follow the input signals is also related to the use of high gain feedback, as is evident from (4.3.17). If the moduli of all $\gamma_{Qi}(j\omega)$ are large, then the CLTM is approximately I and sinusoidal signals of angular frequency ω will thus be accurately tracked. Dynamic tracking, and speed of response, is thus determined by the frequency band over which $Q(j\omega)$ has a relatively large gain.

Steady-state tracking, however, depends only on the gains at $s = 0$. If integral action is incorporated into the compensated system so that $Q(s)$ is of the form $\frac{1}{s}Q_1(s)$ where $Q_1(s)$ has no zeros at $s = 0$, then $\forall$ i

$$|\gamma_{Qi}(s)| \to \infty \qquad \text{as} \qquad s \to 0$$

and the CLTM $\to$ I (see (4.3.17)) giving zero steady-state error to a step input.

Disturbance rejection also requires the use of high-gain feedback. Since (see (4.3.7))

$$y_d(s) = F(s)^{-1}d(s) = (I+Q(s))^{-1}d(s)$$

it follows that, when $Q(s)$ has the form

$$Q(s) = Z(s)\,\mathrm{diag}(\gamma_{Qi}(s))_{i=1}^{m} Z(s)^*$$

$F(s)^{-1}$ will have the form

$$F(s)^{-1} = Z(s)\,\text{diag}\left(\frac{1}{1+\gamma_{Qi}(s)}\right)_{i=1}^{m} Z(s)^* \qquad (4.3.18)$$

and so $F(s)^{-1} \to 0$ if and only if the gains $\gamma_{Qi}(s)$ are all large. In this sense high feedback loop gains will give both good tracking (see (4.3.5)) and good disturbance rejection (see (4.3.7)), since

$$L(s)^{-1} \to I \quad \Leftrightarrow \quad F(s)^{-1} \to 0$$

§4.4 Robustness

By robust stability we broadly mean the ability of $Q(s)$ to remain stable in the face of uncertainties including model inaccuracies, parameter variations and loop gain/phase changes. Many results are emerging in this area (e.g. see [SAN],[LAU],[NUZ],[POS2],[SAF],[CRU],[DOY],[LEH],[SMI2]). In our context, robustness involves a combination of "good stability margins" in the characteristic gain loci with an insensitivity of these loci to change under plant perturbations. It is for a reduction in sensitivity that one seeks to normalize loop transfer functions. We first state some standard results in the general case, and then specialize to normal systems. In particular, we shall interpret the robustness of a normal system in terms of its characteristic gain loci.

Assume that $Q(s) \in \mathbb{R}(s)^{m\times m}$ is closed-loop stable, and that the above uncertainties are taken into account by a multiplicative factor so that the "uncertain system" is $Q(s)\left(I+\Delta(s)\right)$ where we shall further assume that $\Delta(s)$ is stable. Note that if an "additive" perturbation $\delta(s)$ is considered so that the "uncertain system" is $\left(Q(s)+\delta(s)\right)$, then the "additive" perturbation $\delta(s)$ and the "multiplicative" perturbation $\left(I+\Delta(s)\right)$ are related through

$$\Delta(s) = Q(s)^{-1}\,\delta(s)$$

We wish to find a real-valued (positive) function $b(s)$ s.t.

$$\|\Delta(s)\|_2 < b(s) \quad \forall s \in D_{NYQ} \Rightarrow Q(s)\big(I + \Delta(s)\big) \text{ is closed-loop stable} \qquad (4.4.1)$$

b(s) serves as a stability margin (see [SAF]) in the sense that it is a frequency dependent bound for allowable uncertainties $\Delta(s)$ so that any $\Delta(s)$ satisfying the bounding condition will not upset stability. An explicit candidate for b(s) is (see [SAN, Theorem 3] or [SAF, Theorem 3])

$$b(s) = \sigma_{min}(I + Q(s)^{-1}) \qquad s \in D_{NYQ} \qquad (4.4.2)$$

where $\sigma_{min}(\cdot)$ denotes the minimum singular value. Further, if we define

$$b_{min} := \inf_{s \in D_{NYQ}} b(s) \qquad (4.4.3)$$

then b_{min} can be related to a multivariable gain/phase-margin. The following definition (see [SAF], [LEH]) is a generalization of the classical gain/phase-margin concepts to the multivariable case.

Definition 4.4.1

Let Q(s) be closed-loop stable i.e. $(I + Q(s))^{-1}Q(s)$ is stable.

(1) Let $D_G = \text{diag}(d_1, \ldots, d_m)$, $d_i \in \mathbb{R}$. Then the gain margin interval (GMI) is defined to be the largest interval (a,b), $a \leq 1 \leq b$, s.t.

$$d_i \in (a,b) \; \forall i \quad \Rightarrow \quad Q(s) D_G \text{ is closed-loop stable}$$

(2) Let $D_P = \text{diag}(e^{j\theta_1}, \ldots, e^{j\theta_m})$, $\theta_i \in (-\pi, \pi]$. Then the phase margin interval (PMI) is defined to be the largest interval (α, β), $\alpha \leq 0 \leq \beta$, s.t.

$$\theta_i \in (\alpha, \beta) \; \forall i \quad \Rightarrow \quad Q(s) D_P \text{ is closed-loop stable}$$

In other words, GMI (resp. PMI) specifies the limits of allowable pure gain (resp. pure phase) variations in individual loops which could be imposed (possibly simultaneously) without inducing instability. As a direct consequence of (4.4.1), we have

Prop 4.4.2

Let $Q(s)$ be closed-loop stable, and let b_{min} be defined by (4.4.2) and (4.4.3). Then

(1) $(1-b_{min}, 1+b_{min}) \subseteq$ GMI

(2) If $b_{min} < 2$, then $(-2\sin^{-1}\frac{1}{2}b_{min}, 2\sin^{-1}\frac{1}{2}b_{min}) \subseteq$ PMI

If $b_{min} > 2$, then PMI $= (-\pi, \pi]$

Proof :

(1) Let

$$D_G = \text{diag}(d_1, \ldots, d_m) \quad \text{with} \quad d_i \in (1-b_{min}, 1+b_{min}) \ \forall i$$

Now D_G can be written as $(I+\Delta)$ where

$$\Delta = \text{diag}(\delta_1, \ldots, \delta_m) \quad \text{with} \quad \delta_i \in (-b_{min}, b_{min}) \quad \forall i$$

Hence $\|\Delta\|_2 < b_{min}$ and it follows from (4.4.1) that $Q(s)D_G = Q(s)(I+\Delta)$ is closed-loop stable. By Definition 4.4.1, the GMI will at least include $(1-b_{min}, 1+b_{min})$.

(2) Let

$$D_P = \text{diag}(e^{j\theta_1}, \ldots, e^{j\theta_m})$$

and write D_P as $(I+\Delta)$ where

$$\begin{aligned}\Delta &= \text{diag}(e^{j\theta_i} - 1) \\ &= \text{diag}((\cos\theta_i - 1) + j\sin\theta_i) \\ &= 2j\,\text{diag}(e^{j\theta_i/2}\sin\frac{\theta_i}{2})\end{aligned}$$

If $b_{min} < 2$ and $\theta_i \in (-2\sin^{-1}\frac{1}{2}b_{min}, 2\sin^{-1}\frac{1}{2}b_{min})\ \forall\ i$, then

$$2\left|\sin\frac{\theta_i}{2}\right| < b_{min} \tag{4.4.4}$$

If $b_{min} > 2$ then (4.4.4) is satisfied $\forall\ \theta_i \in (-\pi, \pi]$. In both cases $\|\Delta\|_2 < b_{min}$ and the proof proceeds as in (1) above □

In passing, we note that in the usual case when $Q(s)$ is large at $s=0$ and vanishes as $s \to \infty$, we have

$$b(s) \approx \sigma_{min}(I) = 1 \qquad \text{at } s=0$$

and
$$b(s) \approx \sigma_{min}(Q(s)^{-1}) \to \infty \qquad \text{as } s \to \infty$$

However, b_{min} is more often determined by the behaviour of $b(j\omega)$ over the critical frequency range. Typical examples of $b(j\omega)$ will be given in the design studies of Chapters 5 through 7.

Now we specialize to systems which are normal on D_{NYQ}. If

$$Q(s) = Z(s)\Gamma_Q(s)Z(s)^* \qquad \forall\, s \in D_{NYQ}$$

then (4.4.2) can be written

$$\begin{aligned} b(s) &= \sigma_{min}\left(I + Z(s)\Gamma_Q(s)^{-1}Z(s)^*\right) \\ &= \sigma_{min}\left(I + \Gamma_Q(s)^{-1}\right) \\ &= \min_{1\leq i\leq m} \left| \frac{1+\gamma_{Qi}(s)}{\gamma_{Qi}(s)} \right| \end{aligned}$$

A graphical way of determining $b(s)$ from the set of $\gamma_{Qi}(s)$-loci is as follows.

Let $z = x + jy \in \mathbb{C}$ and consider the contour defined by

$$\left| \frac{1+z}{z} \right| = b \qquad (4.4.5)$$

where $b(>0) \in \mathbb{R}$ is some constant. In fact, (4.4.5) defines the familiar M-circles for SISO systems, with $1/M$ here replaced by b. After some rearrangement, (4.4.5) can be written in the form

$$\left(x + \frac{1}{1-b^2}\right)^2 + y^2 = \left(\frac{b}{1-b^2}\right)^2 \qquad (4.4.6)$$

(4.4.6) defines a family of circles parametrized by b, as shown in Fig.4.2(a). We shall call these circles reciprocal M-circles.

Consider a trivial example with

$$Q(s) = \operatorname{diag}\left(\frac{s+3}{s(s+0.4)}, \frac{100}{s(s+10)}\right)$$

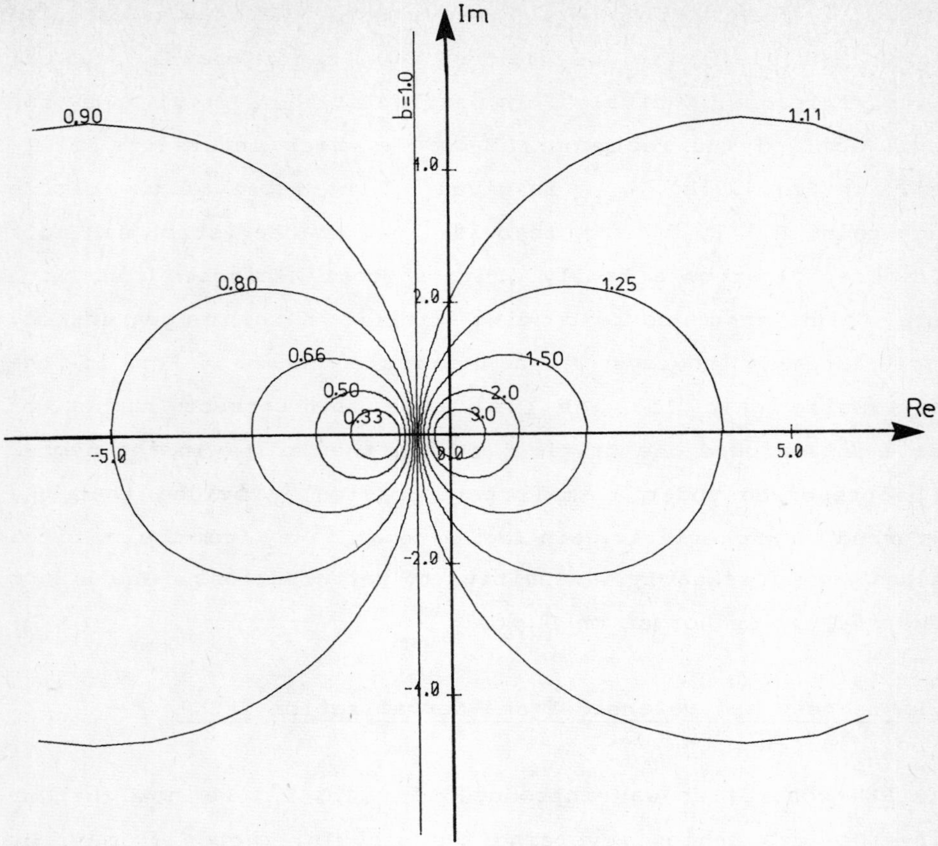

Fig.4.2(a) Reciprocal M-circles.

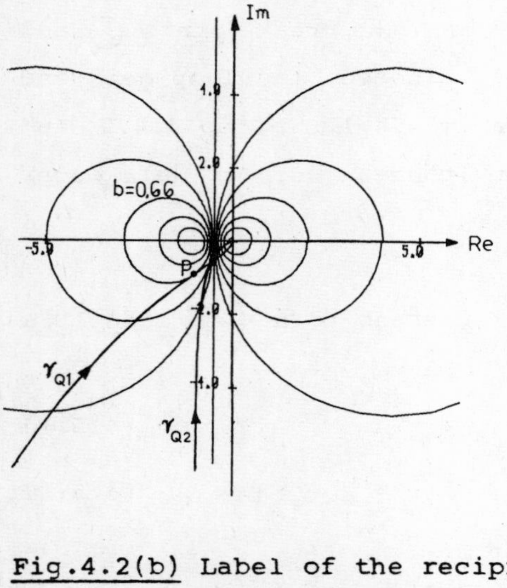

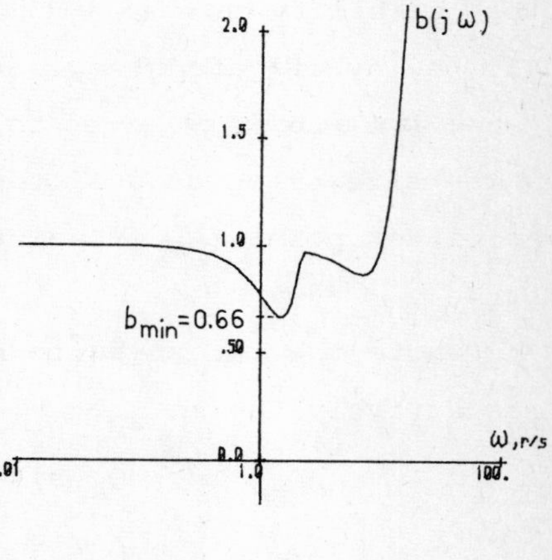

Fig.4.2(b) Label of the reciprocal M-circle through P gives b_{min} of the normal system.

$\gamma_{Q1}(s)$, $\gamma_{Q2}(s)$ are just the diagonal elements. If they are plotted on Fig.4.2(a), giving Fig.4.2(b), then $b(j\omega)$ can be readily read off from the reciprocal M-circles. In particular, b_{min} is given by the minimum label of the reciprocal M-circle which intersects some $\gamma_{Qi}(s)$. In Fig.4.2(b), b_{min} is given by the label of the circle through point P. It follows that if the characteristic gain loci of Q(s) keep clear of a family of reciprocal M-circles round the critical point, then a corresponding stability margin is guaranteed. This is actually obvious in a heuristic sense. For if the characteristic gain loci of Q(s) have the correct number of encirclements around the critical point, then this winding number will be preserved under a small perturbation, provided that the unperturbed characteristic gain loci were well away from the critical point and were reasonably insensitive to perturbations. The latter is true if Q(s) is normal on D_{NYQ}.

§4.5 Robustness and Reversed-Frame-Normalization (RFN)

The RFN controller was introduced in §4.3.1. We now further justify the idea behind reversing the singular-vector frames, by considering the implications for robustness. In a robustness analysis of the closed-loop configuration of Fig.4.3, it is important that stability margins with respect to both the break points Ⓐ and Ⓑ be investigated (e.g. see [POS3]). Indeed, the loop gain and phase variations referred to in Definition 4.4.1 and Prop 4.4.2, and whose effects on closed-loop stability concern us, may well occur at either point Ⓐ or point Ⓑ.

Denote the return-ratio matrices for break points Ⓐ and Ⓑ, respectively, by

$$Q_A(s) := G(s)K(s) \tag{4.5.1}$$

$$Q_B(s) := K(s)G(s) \tag{4.5.2}$$

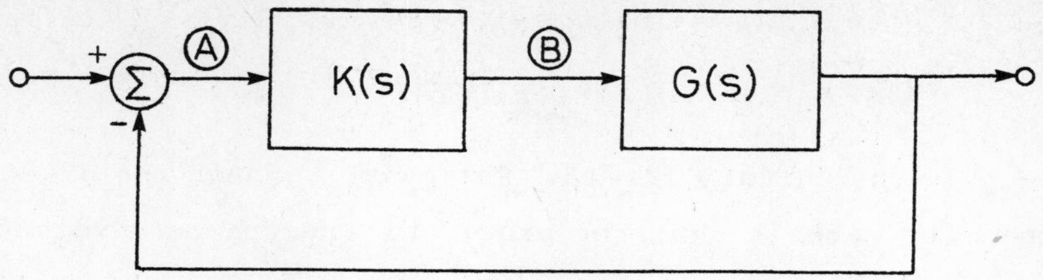

Fig.4.3

Although $Q_A(s)$ and $Q_B(s)$ have the same set of (non-zero) characteristic gain loci, their corresponding robustness measures, namely $\sigma_{min}(I + Q_A(s)^{-1})$ and $\sigma_{min}(I + Q_B(s)^{-1})$ may be quite different. For example, such a discrepancy may arise if among $\{Q_A(s)\ ,\ Q_B(s)\}$, one is close to normality while the other is far from normality. A sufficient condition for the closed-loop system to be robustly stable with respect to perturbations at either points (A) or (B) is that $Q_A(s)$, $Q_B(s)$ are both normal on D_{NYQ} (with good stability margin at either break point). This in turn implies that the singular-vector frames of K(s) must be those of G(s) taken in reversed order, as stated in the following proposition.

Prop 4.5.1

Suppose $G \in \mathbb{C}^{m\times \ell}$ and $K \in \mathbb{C}^{\ell \times m}$ are of full rank r (=min(m,ℓ)) and let

$$G = Z\Gamma_G U^* \tag{4.5.3}$$

where $\Gamma_G \in \mathbb{C}^{r\times r}$ is diagonal, $U \in \mathbb{C}^{r\times \ell}$ and $Z \in \mathbb{C}^{m\times r}$ are subunitary. Then GK and KG are both normal of rank r
iff $K = U\Gamma_K Z^*$ for some diagonal, nonsingular $\Gamma_K \in \mathbb{C}^{r\times r}$. □

A proof of this proposition is given in Appendix B.

Now let G(s) be given by (4.3.12) and let K(s) be an RFN controller of the form (4.3.13). Then

$$Q_A(s) = Z(s)\Gamma_Q(s)Z(s)^*$$

$$Q_B(s) = U(s)\Gamma_Q(s)U(s)^*$$

differ only in the unitary frames. Following through the discussion of §4.4 readily reveals that the exact unitary frames involved have no bearing on any of the results. Hence for a system compensated by an RFN controller, robustness at plant output implies robustness at plant input and vice versa. This is one of the principal justifications for using an RFN controller.

§4.6 Compatibility Conditions

Since Q(s) and G(s) are related by $Q(s) = G(s)K(s)$, it should be observed, when manipulating the characteristic gain loci of Q(s), that certain conditions imposed by the nature of G(s) must not be violated or else we will be aiming at an unattainable target. Some essential rules are given in this section but the list is by no means exhaustive. We shall make the following assumption:

Assumption 4.6.1

The precompensator K(s) has

(1) no poles at $s = \infty$, i.e. K(s) is proper;

(2) no zeros at $s = \infty$, i.e. $\lim_{s\to\infty} K(s)$ is full rank.

A K(s) satisfying Assumption 4.6.1 is said to be <u>regular at ∞</u>. Under this assumption, Q(s) must satisfy the following conditions (C1), (C2) and (C3) compatible with G(s):

(C1) Infinite Zero Structure and Roll-off Rates

As a direct consequence of Assumption 4.6.1, the infinite zero structures of Q(s) and G(s) are identical (e.g. see [VER],[HUN]). In terms of the characteristic gain loci, this implies that the sets of characteristic gain loci of Q(s) and G(s) have the same roll-off rates (provided some mild generic conditions are satisfied).

(C2) Number of Encirclements of the Origin

Let $\gamma_Q \circ D_{NYQ}$ denote the combined characteristic gain loci of $\{\gamma_{Qi} \circ D_{NYQ}\}_{i=1}^{m}$. Then as s goes round the Nyquist D-contour, $\gamma_Q \circ D_{NYQ}$ encircles the origin a number of times given by the difference between the closed RHP zeros and poles of Q(s).

$$\begin{aligned} \#E(\gamma_Q \circ D_{NYQ}, 0) &= \#E((\det Q) \circ D_{NYQ}, 0) \\ &= \#P(\det Q(s), \mathbb{C}_+) - \#Z(\det Q(s), \mathbb{C}_+) \\ &= \#SMP(Q(s), \mathbb{C}_+) - \#SMZ(Q(s), \mathbb{C}_+) \qquad (4.6.1) \end{aligned}$$

Note that the large semi-circle of D_{NYQ} should have a radius R large enough to include all RHP poles and zeros of Q(s). However, in order to consider encirclements of the origin, the semi-circle should also be finite so that $\gamma_Q \circ D_{NYQ}$ will not pass right through the origin.

Now if we require that K(s) has no zeros or poles in the closed RHP except possibly at the origin (i.e. it is stable and minimal phase except possibly with integrators), then all Smith-McMillan poles and zeros of Q(s) in $\mathbb{C}_+^*$ ($:= \mathbb{C}_+ - \{0\}$) are due to G(s), and hence (4.6.1) becomes

$$\begin{aligned} \#E(\gamma_Q \circ D_{NYQ}, 0) = \; &\#SMP(G(s), \mathbb{C}_+) - \#SMZ(G(s), \mathbb{C}_+) \\ &+ \#SMP(K(s), 0) \qquad (4.6.2) \end{aligned}$$

This implies that if a stable, minimal phase compensator is to be constructed, then the total number of origin-encirclements of the final set of characteristic gain loci is prescribed by the number of closed RHP poles and zeros of G(s) plus the number of integrators in the compensator. Violation of (4.6.2) means that at least one closed RHP pole or zero is introduced into K(s).

(C3) Total Phase Change

For the set of characteristic gains $\{\gamma_{Qi} \circ D_{NYQ}\}_{i=1}^{m}$ of Q(s), let

$\Delta_{j\delta}^{jR}$ arg $\gamma_{Qi}(s)$ denote the net phase change of the loci of $\gamma_{Qi}(s)$ as s goes up the imaginary axis, along D_{NYQ}, between $j\delta$ and jR (see Fig.4.4). Define the <u>total phase change</u> of the set of characteristic gain loci of Q(s) by

$$TPC(Q(s)) := \sum_{i=1}^{m} \lim_{\substack{\delta \to 0 \\ R \to \infty}} \Delta_{j\delta}^{jR} \arg \gamma_{Qi}(s)$$

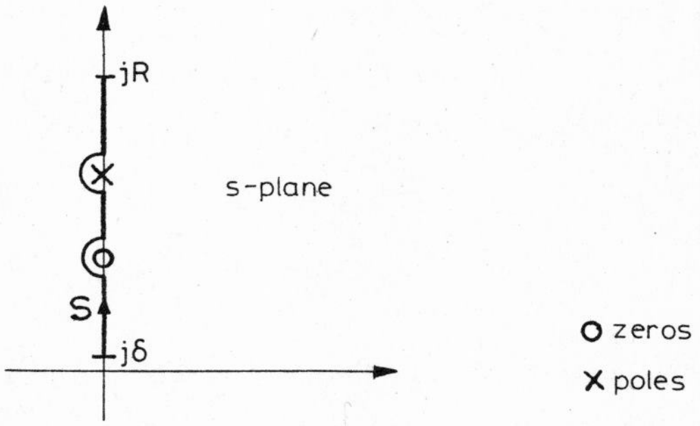

Fig.4.4

The following theorem [SMI2] is a multivariable generalization of a result of Bode about the net phase change of a scalar transfer function.

Theorem 4.6.2

Let $Q(s) = G(s)K(s)$ where $Q(s), G(s), K(s) \in \mathbb{R}(s)^{m\times m}$ have full rank. Then:

(1) $$TPC(Q(s)) = [2(P_R - Z_R) + (P_o - Z_o) - \#IZ(Q(s))] \cdot \frac{\pi}{2}$$

where $P_R - Z_R := \#SMP(Q(s), \mathbb{C}_+^*) - \#SMZ(Q(s), \mathbb{C}_+^*)$

$P_o - Z_o := \#SMP(Q(s), 0) - \#SMZ(Q(s), 0)$

(2) $$TPC(Q(s)) = TPC(G(s)) + TPC(K(s))$$

(3) If K(s) satisfies Assumption 4.6.1 and has

(i) no zeros in $\mathbb{C}_+$ (i.e. minimal phase)

(ii) no poles in $\mathbb{C}_+^*$ (i.e. stable except possibly with integrators), then

$$TPC(Q(s)) = TPC(G(s)) + \#SMP(K(s), 0) \cdot \frac{\pi}{2} \qquad \square$$

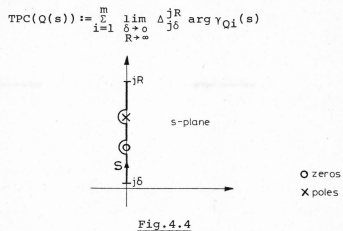

A proof of statement (1) using expansions of algebraic functions can be found in [SMI2]. A slightly modified version of the proof is given in Appendix B. Again the system and the controller, especially if one which is stable and minimal phase is desired, impose specific restrictions on the overall phase variation of the compensated system.

It seems appropriate at this point to explain the reason for going into considerable detail in describing some necessary relationships between the set of characteristic gain loci of Q(s) and G(s). Suppose a precompensator K(s) has been synthesized. If we then put Q(s) = G(s)K(s), clearly the above three conditions will, a posteriori, be automatically satisfied and so it is meaningless to check them after a design has been done. However, since the approach to be taken is to prescribe a certain desirable Q(s) and then come back to see if it is possible to construct a corresponding K(s), the compatibility conditions are thus important as a preliminary check when prescribing Q(s).

Finally, we remark that conditions (C1), (C2) and (C3) are in fact dependent in that any two of them imply the third. But it is always good practice, as an extra check, to ensure that all three conditions are satisfied.

§4.7 Specifying a Desired Compensated System

With the ideas developed so far, we can now summarize what we would aim at as an ideally compensated hypothetical system.

Given a system G(s), the precompensator K(s) should, for robustness reasons, have the same singular-vector frames as G(s), taken in reversed order. That is, the ideally compensated system Q(s) = G(s)K(s) takes the form (4.3.14). Q(s) will then be completely determined once its characteristic gains $\Gamma_Q(s) = \mathrm{diag}(\gamma_{Qi}(s))$ are specified. The characteristic gains $\gamma_{Qi}(s)$ should, of course, be chosen with the requirements for stability, performance and robustness in mind. Stability is simply a matter of ensuring that $\gamma_Q \circ D_{NYQ}$ has the correct number of encirclements around the critical point. Performance, in general terms, can be stipulated as involving integral action, a suitably high gain over a sufficiently wide operating bandwidth, and diagonalization or gain/phase-balancing around the cross-over frequency. Robustness requires, in addition to the reversed frames, that the $\gamma_{Qi}(j\omega)$'s stay well away from the critical point. In addition to all these desirable features, the set of $\gamma_{Qi}(j\omega)$'s must be consistent with the compatibility conditions imposed by G(s).

More often than not, all the above requirements still leave the designer a large class of $\{\gamma_{Qi}(s)\}$ to choose from. It is at this point that physical aspects, such as allowable loop gains, power considerations, plant input-saturation-level constraints etc., come into consideration and provide further guidance for the design.

An example is now given to illustrate these general ideas.

Example 4.7.1

Consider the system G(s) for AUTM whose QN loci have been shown in Example 3.5.1. Given the output frame Z(s) of the QN decomposition $G(S) = Z(s)\Gamma_G(s)U(s)^*$, then using the RFN design approach (see §4.3.1), the compensated system Q(s) will have the form $Z(s)\Gamma_Q(s)Z(s)^*$. We now wish to specify a $\Gamma_Q(s)$ which will have the desirable properties stated in §4.2 through §4.5 and satisfy the conditions of §4.6. A plausible simple candidate (clearly there are many others) is

$$\Gamma_Q(s) = \mathrm{diag}(\gamma_{Q1}(s), \gamma_{Q2}(s))$$

$$= \mathrm{diag}\left(\frac{5}{s}, \frac{50}{s(s+10)}\right) \qquad (4.7.1)$$

The Nyquist and Bode plots of $\gamma_{Q1}(j\omega)$, $\gamma_{Q2}(j\omega)$ are given in Fig.4.5(a,b). The reasons for the choice (4.7.1) are:

(1) Integral action has been incorporated into $\gamma_{Q1}(s)$, $\gamma_{Q2}(s)$ and

hence the steady-state error to a step response will be zero. Furthermore, the high gains at low frequencies and well-balanced characteristic gains up to the cross-over region will ensure low interaction when the loop is closed round Q(s).

(2) G(s) is open-loop stable while the precompensator K(s) is expected to contain 2 poles at the origin to provide the integral action. Hence, for closed-loop stability, $\gamma_{Q1} \circ D_{NYQ}$, $\gamma_{Q2} \circ D_{NYQ}$ should encircle the critical point -1 anticlockwise twice.

(3) $\gamma_{Q1}(j\omega)$, $\gamma_{Q2}(j\omega)$ keep well away from the critical point, with $b_{min} = 1$ (see (4.4.2) and (4.4.3)). The GMI and PMI will thus at least include (0,2), $(-\pi/3, \pi/3)$ (see Prop 4.4.2).

To check the compatibility conditions, we redraw Fig.4.5(a) as Fig.4.5(c), which shows the full image curves of D_{NYQ} under the maps $\gamma_{Q1}(s)$ and $\gamma_{Q2}(s)$. Note that

(1) The infinite zeros of $\gamma_{Q1}(s)$, $\gamma_{Q2}(s)$ are of orders 1, 2 respectively. This is consistent with the infinite zero structure of AUTM. (see remark (2)(ii) after Example 2.3.2).

(2) For this example, (4.6.2) reads $\#E(\gamma_Q \circ D_{NYQ}, 0) = 2$, which is the case in Fig.4.5(c).

(3) From Fig.4.5(b),

$$TPC(Q(s)) = -\frac{\pi}{2}$$

Since G(s) has neither poles nor zeros in $\mathbb{C}_+$, using Theorem 4.3.2(1),

$$\begin{aligned} TPC(G(s)) &= (2\times 0 + 0 - (1+2))\cdot\frac{\pi}{2} \\ &= -\frac{3\pi}{2} \end{aligned}$$

If the precompensator K(s) is to have no poles or zeros in the closed RHP except possibly with integrators, then Theorem 4.6.2(3) gives

$$\#SMP(K(s),0) = 2 \qquad (4.7.2)$$

i.e. K(s) should have two poles at s = 0, which is expected because the integral action injected into $\gamma_{Q1}(s)$, $\gamma_{Q2}(s)$ must come from the precompensator K(s).

Hence as far as the compatibility conditions are concerned, there are no contradictions.

So far we have not addressed the important question of whether it is possible to construct the required K(s), and if so how. This now will be examined in the following two chapters.

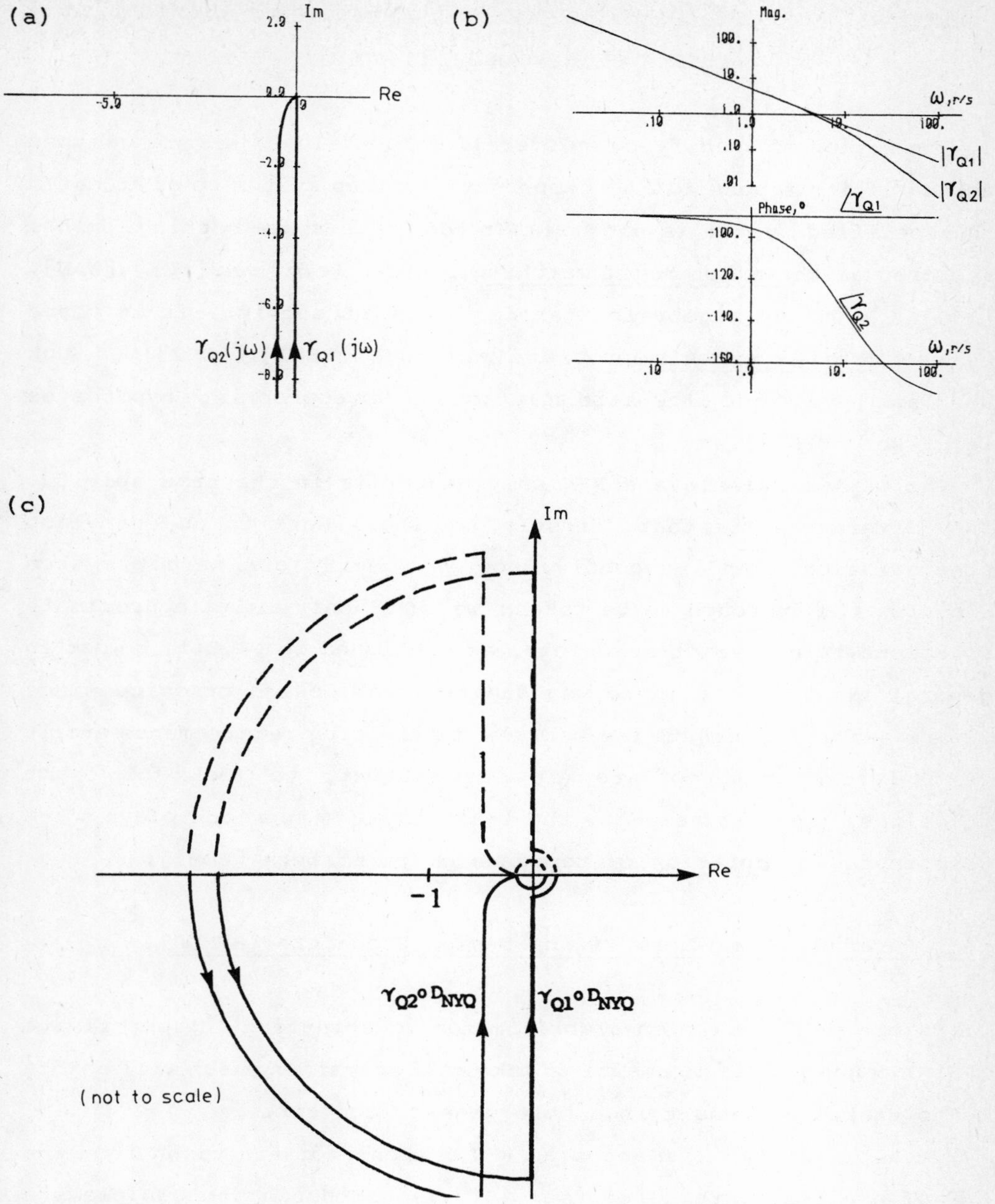

Fig.4.5

(a),(b) The specified set of $\{\gamma_{Q1}(j\omega), \gamma_{Q2}(j\omega)\}$.

(c) Full image curves of D_{NYQ} under γ_{Q1}, γ_{Q2} show two encirclements of the origin and two encirclements of the critical point.

CHAPTER 5 CALCULATING A COMPENSATOR NUMERATOR MATRIX BY LINEAR LEAST-SQUARES FITTING

The notion of specifying an overall (closed-loop) system and then asking if it is possible to compensate a given system so as to match the specified system is not new. In the state-space setting, it has appeared as the <u>exact model matching</u> problem (e.g. see [WOL],[WAN], [MOO]). In the algebraic transfer function setting, it is posed as the <u>total synthesis</u> problem (e.g. see [PEC1],[PEC2]). Such problems have exact algebraic solutions under appropriate hypotheses (e.g. see [DES2]).

The way we calculate a RFN compensator is in the same spirit. The differences are that, firstly, we shall work in an open-loop configuration. And, secondly, how we specify our target system dictates the approach to be taken: we shall only aim at approximate solutions, since exact solutions, as mentioned in §4.3.1, do not in general exist. This in no way departs from design practice since we are seeking a compromise between conflicting requirements and it is therefore inappropriate, in our context, to look for exact solutions. In this respect, the least-squares approach plays a key role in the calculation of compensator parameters [EDM1].

§5.1 Reversed-Frame-Normalizing Design Procedure (RFNDP)

Before we describe an algorithm for constructing an approximate RFN compensator, it is useful to make an observation which will enable us to state the result in a more general context.

Recall that the frames of an RFN compensator are taken from the QND of a given system $G(s)$ (see §4.3.1). Although the QND plays a key role in the conceptual framework of the RFN approach, a closer examination of the ideally compensated system $Q(s)$ (see (4.3.14)) reveals that it is actually independent of whether a QND or an SVD is used. Suppose $G(s)$ has a QND and an SVD given by

$$G(s) = Z(s)\Gamma_G(s)U(s)^* \qquad (QND) \qquad (5.1.1)$$

$$= Y(s)\Sigma_G(s)U(s)^* \qquad (SVD) \qquad (5.1.2)$$

Then Z(s) and Y(s) are related by (see(3.3.6) and the remark following (3.3.7))

$$Z(s) = Y(s)\Theta(s)^*$$

for some diagonal unitary $\Theta(s)$. Hence the Q(s) of (4.3.14) can be written:

$$Q(s) = Z(s)\Gamma_Q(s)Z(s)^* \qquad (5.1.3)$$

$$= Y(s)\Theta(s)^*\Gamma_Q(s)\Theta(s)Y(s)^*$$

$$= Y(s)\Gamma_Q(s)Y(s)^* \qquad (5.1.4)$$

Computationally, (5.1.4) is to be preferred over (5.1.3) because Y(s) is readily obtained by applying a standard SVD routine to G(s).

The next algorithm to be discussed is a direct implementation of the ideas of the RFN compensator by an optimization technique. The algorithm is stated for the case of designing a precompensator. A similar version for postcompensator design is obtained in an obvious way.

Algorithm 5.1.1 (RFNDP: Precompensator Numerator Fitting)

(1) Choose a frequency list $\{\omega_1,\ldots,\omega_\rho,\ldots,\omega_n\}$. For each frequency ω_ρ, do an SVD of the given system $G(s) \in \mathbb{R}(s)^{m\times \ell}$:

$$G_\rho := G(j\omega_\rho) = Y_\rho \Sigma_{G\rho} U_\rho^* \qquad (5.1.5)$$

where Y_ρ, $\Sigma_{G\rho}$, U_ρ^* are respectively $m\times r$, $r\times r$, $r\times \ell$ and $r := \min(m,\ell)$.

(2) Specify a set of desired characteristic gain loci $\{\gamma_{Qi}(s)\}_{i=1}^{r}$ for the desired precompensated system. At each frequency ω_ρ, calculate the desired precompensated system by

$$Q_\rho := Y_\rho \Gamma_Q(j\omega_\rho) Y_\rho^* \qquad (5.1.6)$$

where $\Gamma_Q(s) := \mathrm{diag}(\gamma_{Q1}(s),\ldots,\gamma_{Qr}(s))$

(3) Choose a polynomial matrix $D(s) \in \mathbb{R}[s]^{\ell\times \ell}$ for the denominator

of a left matrix fraction description (MFD) for the precompensator

$$K(s) = D(s)^{-1}N(s)$$

where $N(s) \in \mathbb{R}[s]^{\ell \times m}$ is a numerator polynomial matrix whose coefficients are to be determined in step (5) below. At each frequency ω_ρ, evaluate

$$D_\rho := D(j\omega_\rho)$$

(4) Choose a weighting matrix $W(s) \in \mathbb{R}(s)^{m \times m}$ and let $W_\rho := W(j\omega_\rho)$.

(5) Define the error matrix E_ρ to be

$$E_\rho := G_\rho D_\rho^{-1} N(j\omega_\rho) - Q_\rho \qquad \rho = 1, \ldots, n \qquad (5.1.7)$$

and determine the coefficients of $N(s)$ by solving the problem:

$$\text{minimize} \quad \sum_{\rho=1}^{n} \| E_\rho \|_{W_\rho}^2 \qquad (5.1.8)$$

over some specified parameter space of $N(s)$. (How the parameter space is specified will be explained in §5.3.)

End of Algorithm 5.1.1

Notice that Algorithm 5.1.1 applies to systems with an arbitary number of inputs and outputs. However some of the results of Chapter 4 may not have a direct interpretation if the system $G(s)$ has a different number of inputs and outputs. We defer the discussion of such issues, for non-square systems, to the last chapter. For the benefit of the discussion to be given there, we emphasize that Algorithm 5.1.1, regarded as a fitting algorithm, is not restricted to square systems.

In step (5), we are minimizing over a finite frequency list the sum of squares of the differences between two sets of Nyquist arrays — for the system $G(s)$ precompensated by a yet undetermined precompensator and for a desired response defined in step (2). This means that the numerator of $K(s)$ is synthesized so that $G(s)K(s)$

will fit the desired response as closely as possible in a least-square sense. This minimization problem can be put into the form of a linear least-squares problem. Before doing so, we digress to review some of the standard results that will be needed.

§5.2 Some Results for the Linear Least-Squares Problem

The relevant problem of minimizing a sum of squares can be stated as:

$$\begin{cases} \text{Given} \quad A \in \mathbb{R}^{t\times s}, \quad q \in \mathbb{R}^t \\ \underset{n \in \mathbb{R}^s}{\text{minimize}} \; \| An - q \|^2 \end{cases} \qquad (5.2.1)$$

Theorem 5.2.1

Any solution $\hat{n} \in \mathbb{R}^s$ to the linear least squares problem (5.2.1) satisfies the normal equation:

$$(A^T A)\hat{n} = A^T q \qquad (5.2.2)$$

If A has full rank, then the solution is unique and is given by

$$\begin{aligned} \hat{n} &= (A^T A)^{-1} A^T q \\ &= A^{\dagger} q \end{aligned} \qquad (5.2.3)$$

where $A^{\dagger}$ is the Moore-Penrose inverse of A. □

A proof of this elementary result can be found in any standard text (e.g. see [BEN, chapter 3 §1]). Numerical considerations, as well as extensive software for solving the linear least-squares problem, are given in [HAN],[DON, chapters 9 & 11].

Geometrically, $A\hat{n}$ ($= AA^{\dagger}q$, see(5.2.3)) is the orthogonal projection of q onto the range space of columns of A. Fig.5.1 depicts the ideas for the case t=3, s=2.

The residual sum of squares is given by $\| q - A\hat{n} \|^2$ and we shall use the relative error

$$\frac{\| q - A\hat{n} \|}{\| q \|} \tag{5.2.4}$$

as an indicator of how good the fitting is.

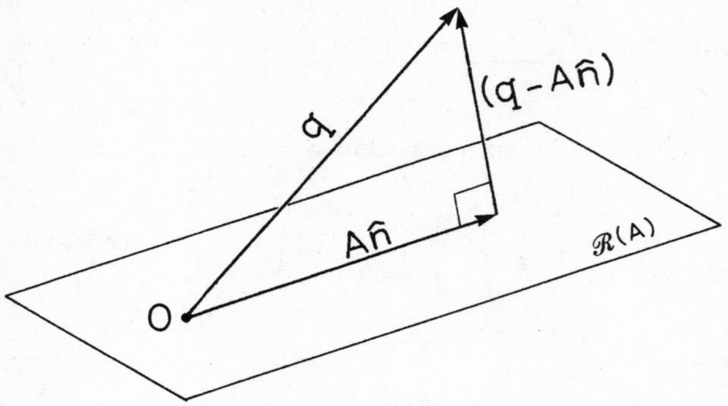

Fig.5.1

If instead of (5.2.1), we are required to

$$\underset{n \in \mathbb{R}^s}{\text{minimize}} \; \| An - q \|_W^2$$

where $W = P^T P$ is a positive definite (weighting) matrix and $\| \cdot \|_W$ denotes the weighted Euclidean vector norm, then the solution (5.2.2) and (5.2.3) should be modified as

$$(A^T W A)\hat{n} = A^T W q$$

If $(A^T W A)$ is nonsingular, then

$$\begin{aligned} \hat{n} &= (A^T W A)^{-1}(A^T W q) \\ &= (PA)^{\dagger}(Pq) \end{aligned} \tag{5.2.5}$$

Now, since we shall be fitting Nyquist arrays, consisting of complex entries, we need to consider the following modified version of the least-squares problem

$$\left\{ \begin{array}{l} \text{Given } \; A \in \mathbb{C}^{t \times s}, \quad q \in \mathbb{C}^t \\ \text{and a positive definite, hermitian } \; W = P^* P \in \mathbb{C}^{t \times t}, \\ \underset{n \in \mathbb{R}^s}{\text{minimize}} \; \| An - q \|_W^2 \end{array} \right. \tag{5.2.6}$$

Corollary 5.2.2

Any solution $\hat{n} \in \mathbb{R}^s$ to (5.2.6) satisfies

$$\mathrm{Re}(A^*WA)\hat{n} = \mathrm{Re}(A^*Wq) \tag{5.2.7}$$

If PA has full rank, then the solution is unique and is given by

$$\begin{aligned}\hat{n} &= [\mathrm{Re}(A^*WA)]^{-1}\mathrm{Re}(A^*Wq)\\ &= \begin{bmatrix}\mathrm{Re}(PA)\\ \mathrm{Im}(PA)\end{bmatrix}^{\dagger}\begin{bmatrix}\mathrm{Re}(Pq)\\ \mathrm{Im}(Pq)\end{bmatrix}\end{aligned} \tag{5.2.8}$$

Proof:

The expression to be minimized can be rewritten as

$$\begin{aligned}\|An - q\|_W^2 &= (An-q)^*W(An-q)\\ &= (PAn-Pq)^*(PAn-Pq)\end{aligned} \tag{5.2.9}$$

Let $x = PAn - Pq$ and note that

$$\begin{aligned}x^*x &= (\mathrm{Re}\,x - j\,\mathrm{Im}\,x)^T(\mathrm{Re}\,x + j\,\mathrm{Im}\,x)\\ &= (\mathrm{Re}\,x)^T(\mathrm{Re}\,x) + (\mathrm{Im}\,x)^T(\mathrm{Im}\,x)\\ &= \left\|\begin{bmatrix}\mathrm{Re}\,x\\ \mathrm{Im}\,x\end{bmatrix}\right\|^2\end{aligned}$$

Putting this into (5.2.9), the expression to be minimized becomes

$$\|An - q\|_W^2 = \left\|\begin{bmatrix}\mathrm{Re}(PA)\\ \mathrm{Im}(PA)\end{bmatrix}n - \begin{bmatrix}\mathrm{Re}(Pq)\\ \mathrm{Im}(Pq)\end{bmatrix}\right\|^2$$

Since all matrices in the norm expression on the right are real we can apply Theorem 5.2.1 to the present problem. In particular, (5.2.8) follows immediately from (5.2.3). □

With these results, we can tackle the problem stated in step (5) of Algorithm 5.1.1.

§5.3 Calculation of the Precompensator Numerator Matrix

In Algorithm 5.1.1, we have chosen to work with a left MFD for the precompensator

$$K(s) = D(s)^{-1}N(s) \quad \varepsilon\ \mathbb{R}(s)^{\ell\times m}$$

where $D(s)\ \varepsilon\ \mathbb{R}[s]^{\ell\times\ell}$, $N(s)\ \varepsilon\ \mathbb{R}[s]^{\ell\times m}$

In order to obtain a proper precompensator, it is necessary to impose conditions on the degrees of the polynomial entries of N(s). Let the row degrees (i.e. maximum degrees of the polynomials in each row) of D(s) be

$$d_i := \deg(\text{row}_i(D(s))) \qquad i = 1,\ldots,\ell$$

where $\text{row}_i(\cdot)$ denotes the ith row of the matrix. Furthermore, suppose that D(s) is row reduced (or row proper, i.e. $\deg(\det D(s)) = \sum_{i=1}^{\ell} d_i$), which is justified because D(s) is chosen by the designer. Then the requirement that K(s) be proper is equivalent to the conditions (e.g. see [KAI, pp.385 Lemma 6.3-11])

$$\deg(\text{row}_i(N(s))) \leqslant d_i \qquad i = 1,\ldots,\ell$$

To allow the maximum number of free parameters in the minimization step, we shall put

$$\deg(n_{ij}(s)) = d_i \qquad i = 1,\ldots,\ell\ ;\ j = 1,\ldots,m$$

where $n_{ij}(s)$ denotes the (i,j)th entry of N(s).

Let

$$n_{ij}(s) = \sum_{k=o}^{d_i} n_{ij}^k\, s^k$$

where n_{ij}^k are the coefficients of $n_{ij}(s)$ to be estimated. We can collect all of the coefficients together, into a coefficient matrix, as follows. Define

$$S(s) := \begin{bmatrix} \begin{array}{cccc:} 1 & s & \dots & s^{d_1} \end{array} & & \bigcirc \\ & \ddots & \\ \bigcirc & & \begin{array}{:cccc} 1 & s & \dots & s^{d_\ell} \end{array} \end{bmatrix} \tag{5.3.1}$$

and

$$N := \begin{bmatrix} n_{11}^0 & n_{11}^1 & \dots & n_{11}^{d_1} & \dots & n_{\ell 1}^0 & n_{\ell 1}^1 & \dots & n_{\ell 1}^{d_\ell} \\ & & \dots & & \dots & & & \dots & \\ n_{1m}^0 & n_{1m}^1 & \dots & n_{1m}^{d_1} & \dots & n_{\ell m}^0 & n_{\ell m}^1 & \dots & n_{\ell m}^{d_\ell} \end{bmatrix}^T \quad \varepsilon \quad \mathbb{R}^{(\ell+\Sigma d_i)\times m}$$

then

$$N(s) = S(s)N \tag{5.3.2}$$

The problem of finding $N(s)$ now reduces to the determination of the coefficient matrix N. Using (5.3.2), the error matrix of (5.1.7) can be written as:

$$\begin{aligned} E_\rho &= G_\rho D_\rho^{-1} S(j\omega_\rho)N - Q_\rho \\ &= A_\rho N - Q_\rho \qquad \rho = 1,\dots,n \end{aligned} \tag{5.3.3}$$

where $\quad A_\rho := G_\rho D_\rho^{-1} S(j\omega_\rho)$

Putting

$$E := \begin{bmatrix} E_1 \\ \vdots \\ E_n \end{bmatrix} \qquad A := \begin{bmatrix} A_1 \\ \vdots \\ A_n \end{bmatrix} \qquad Q := \begin{bmatrix} Q_1 \\ \vdots \\ Q_n \end{bmatrix} \qquad W := \begin{bmatrix} W_1 \\ \vdots \\ W_n \end{bmatrix}$$

the n equations of (5.1.7) (or (5.3.3)) can be stacked together to form a single equation

$$E = AN - Q$$

and accordingly, (5.1.8) becomes

$$\underset{N}{\text{minimize}}\ \|E\|_W^2 \equiv \underset{N}{\text{minimize}}\ \|AN - Q\|_W^2 \tag{5.3.4}$$

This is of the form (5.2.6) except that the column vectors n, q of (5.2.6) are replaced by the matrices N, Q. However, now we can simply apply Corollary 5.2.2 columnwise to (5.3.4) and solve for the complete

N, and hence N(s), in m (= number of columns of N) steps. This completes Algorithm 5.1.1.

§5.4 Example

A few remarks about the execution of Algorithm 5.1.1 are in order. Once implemented on a computer, the work of the designer in executing this algorithm is reduced to the choices to be made in steps (2), (3) and (4).

As far as deciding what characteristic gain loci $\gamma_{Qi}(s)$ to specify, some basic guidelines have already been discussed in Chapter 4. However, much flexibility is still left to the designer and evidently, what is chosen must depend specifically on the given system as well as on engineering insight obtained by a preliminary study of the system. Such an initial study may be done along the lines of §3.6 and §3.7. Although it is not absolutely necessary to require that the $\gamma_{Qi}(s)$'s be rational functions, it seems reasonable to do so because this makes the compatibility condition (C3) easier to handle.

The second choice the designer has to make is that of the denominator matrix D(s), and hence the poles of the precompensator. This choice is often not obvious and so the following simple approach is suggested. Let $D(s) = d(s)I_\ell$ and select a common denominator polynomial d(s) for K(s). The simplicity, of course, is achieved at the expense of losing some degrees of freedom when synthesizing K(s). For this reason we shall proceed, in the next chapter, to nonlinear least-squares techniques which will release the designer from this difficult choice. For the moment, we shall be content with the simpler approach.

Finally, it is usually quite easy to choose an appropriate weighting matrix W(s), as will be illustrated by an example below.

The following example illustrates the reversed-frame-normalizing and linear least-squares fitting procedure.

Example 5.4.1

Consider the system AUTM again (see Appendix C). We shall follow through the steps of Algorithm 5.1.1 to find a precompensator for this system. All the calculations are based on a logarithmic equally spaced frequency list, {0.01,...,100}, of 50 points.

Step (1) is just a direct computation. As to step (2), we shall stick to the choices $\gamma_{Q1}(s)$, $\gamma_{Q2}(s)$ made in Example 4.7.1. The desired characteristic gain loci $\Gamma_Q(j\omega_\rho)$ and the desired compensated Nyquist array Q_ρ, $\rho = 1,\ldots,50$ (see (5.1.6)) are shown in Fig.5.2(a-d). (For simplicity, we have put axes only on the (1,1)-elements of the arrays given in Fig.5.2(c,d). Whenever axes are missing from the other elements, it is to be understood that all entries are drawn to the same scale as that for the (1,1)-element.)

For step (3), we choose, for simplicity, the denominator matrix

$$D(s) = s(s+2)I_2$$

Note that the prescribed set of loci $\gamma_{Q1}(s)$, $\gamma_{Q2}(s)$ presumes that the precompensator has 2 integrators and hence requires the factor s in D(s). The other two poles of the precompensator, at $s = -2$, are chosen to be somewhat faster than the dominant poles of the system.

Next, the weighting matrix is chosen to be

$$W(s) = |w(s)|^2 \mathbb{1}_2 \qquad (5.4.1)$$

where

$$w(s) = \frac{100(s+0.1)}{(s+10)} \qquad (5.4.2)$$

The magnitude plot of $|w(s)|$ is given in Fig.5.2(e). The reason for using a weighting which increases with frequency is to counterbalance the decreasing magnitude of the desired response, for otherwise the least-squares fitting will more or less ignore the medium and high frequency data which are crucial for stability.

Having completed the above four steps, now all the data can be processed by the computer, which performs the calculations described in §5.3, giving the precompensator

$$K(s) = \frac{1}{s(s+2)} \begin{bmatrix} 2.08s^2 + 3.66s + 6.12 & -0.451s^2 - 0.026s - 1.87 \\ -0.260s^2 - 0.731s - 1.47 & 0.531s^2 + 5.19s + 1.23 \end{bmatrix}$$

with a weighted relative error of fitting (see (5.2.4)) = 0.139

It remains to check the actual precompensated system $G(s)K(s)$, which is expected to be different from the desired response since the fitting is not perfect. Fig.5.3 shows the Nyquist array (a,b), the characteristic gain loci (c,d), the QN loci (e,f), the bands around the two branches of the QN loci (g,h) and the frame-misalignment (i) of the precompensated system. These correspond to acceptable closed-loop behaviour. Perhaps slightly surprising is that the frame misalignment does not seem to have improved over the uncompensated system (see Fig.3.3(f)).

In order to check robustness, the functions (see (4.4.2))

$$\sigma_{min}[I + (G(j\omega)K(j\omega))^{-1}] \quad \text{and} \quad \sigma_{min}[I + (K(j\omega)G(j\omega))^{-1}]$$

corresponding to loop-breaking points at the plant output and at the plant input, are plotted in Fig.5.4(a,b). They show that the design has stability margins close to what was specified.

As a matter of interest, we have also shown the PG loci and the frame angles of the precompensated system in Fig.5.5(a-d).

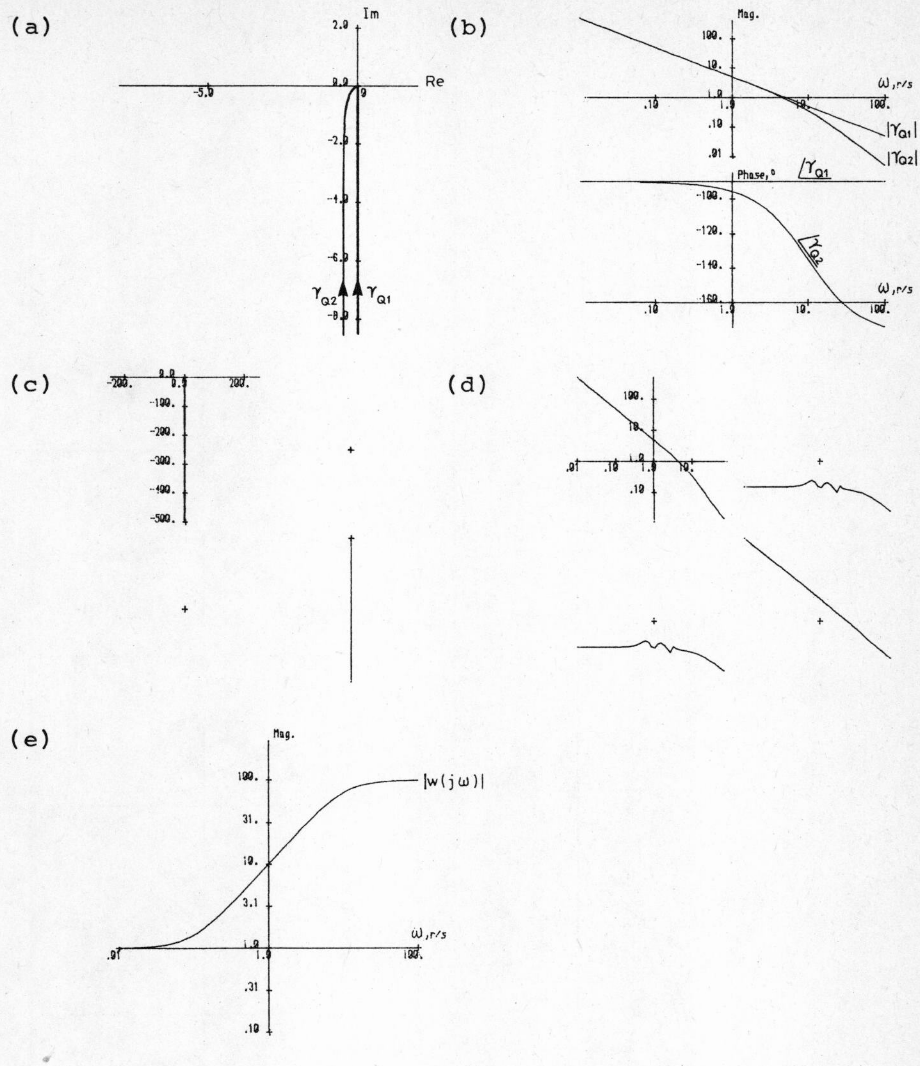

Fig.5.2 Graphs showing choices made in steps (2) and (4) of Algorithm 5.1.1, for the system AUTM.

(a),(b) Specified desired characteristic gain loci:

$$\gamma_{Q1}(s) = \frac{5}{s}\ , \qquad \gamma_{Q2}(s) = \frac{50}{s(s+10)}\ .$$

(c),(d) Nyquist and Bode magnitude arrays of the desired response. All elements are drawn to the same scale as the (1,1) entry. This applies in all subsequent array diagrams.

(e) The weighting function.

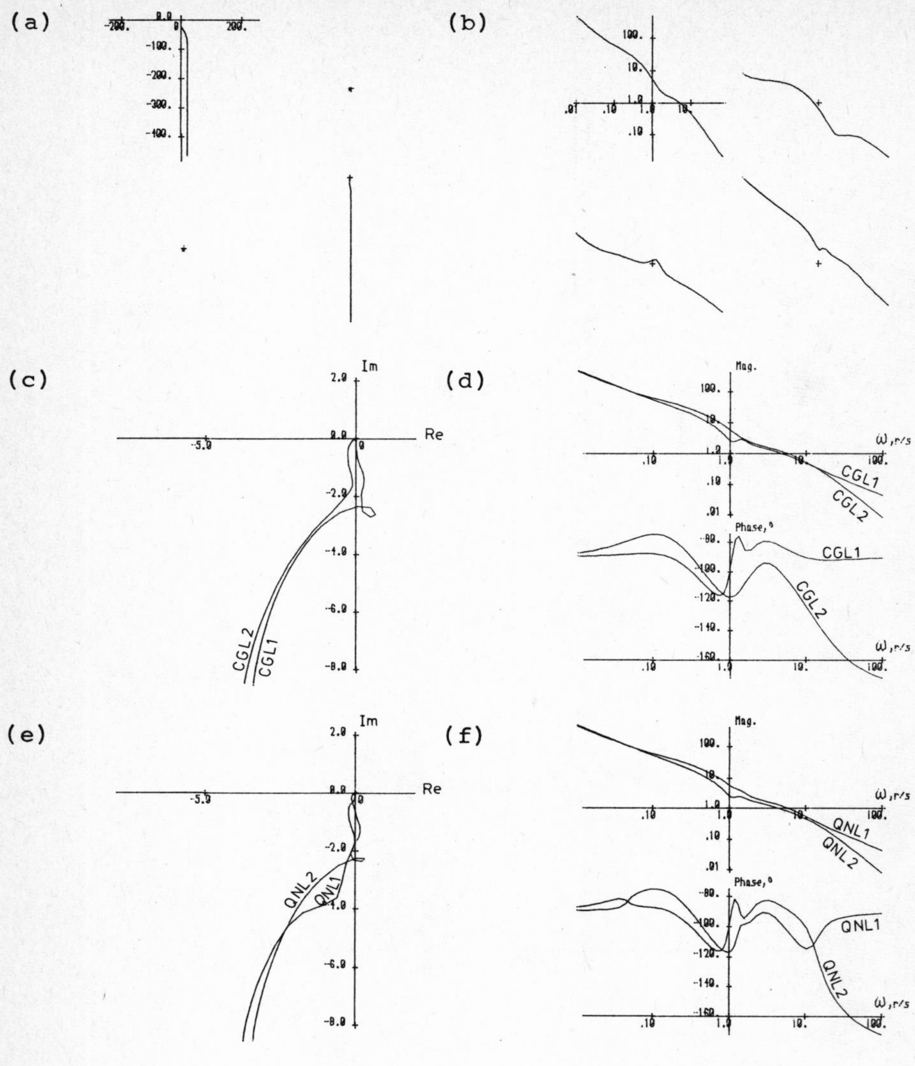

<u>Fig.5.3</u> After the linear least-squares fitting, results for the precompensated system G(s)K(s) are given above.

(a),(b) Nyquist and Bode magnitude arrays (compare with Fig.5.2 (c,d)).

(c),(d) The characteristic gain loci show that G(s)K(s) is closed-loop stable.

(e),(f) QN loci.

(g)

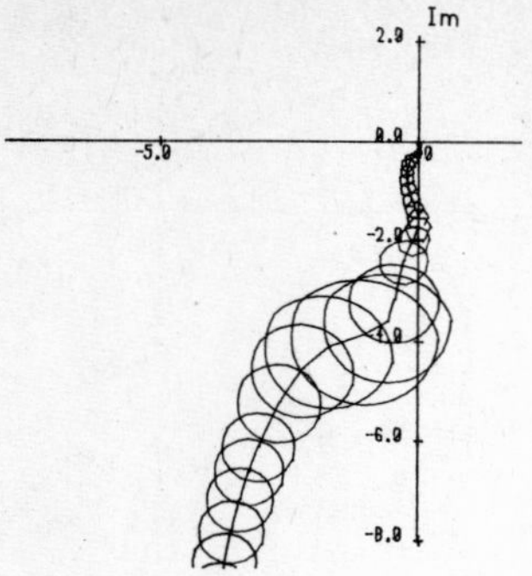

(h)

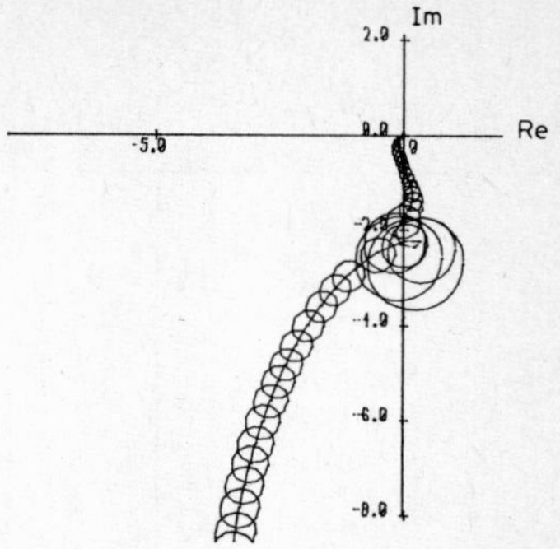

(i)

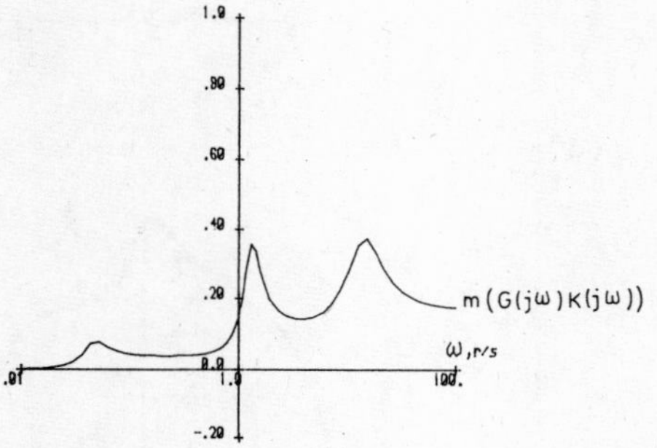

Fig.5.3 (continued)

(g),(h) QNLi taken with the bands $B_i(j\omega)$ (i=1,2).

(i) Frame misalignment (compare with Fig.3.3(f)).

(a)

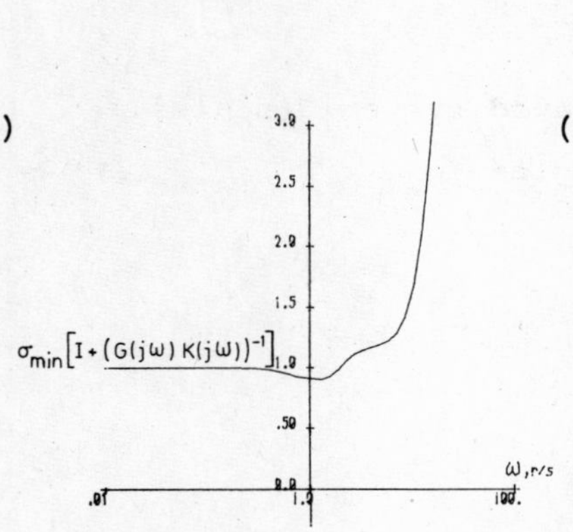

(b)

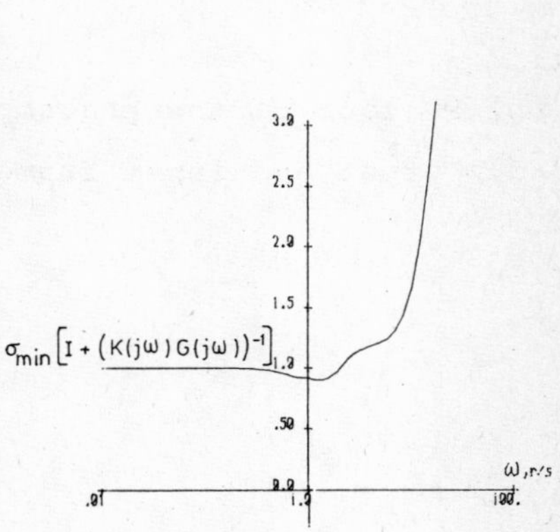

Fig.5.4 Robustness measures at (a) plant output; and (b) plant input.

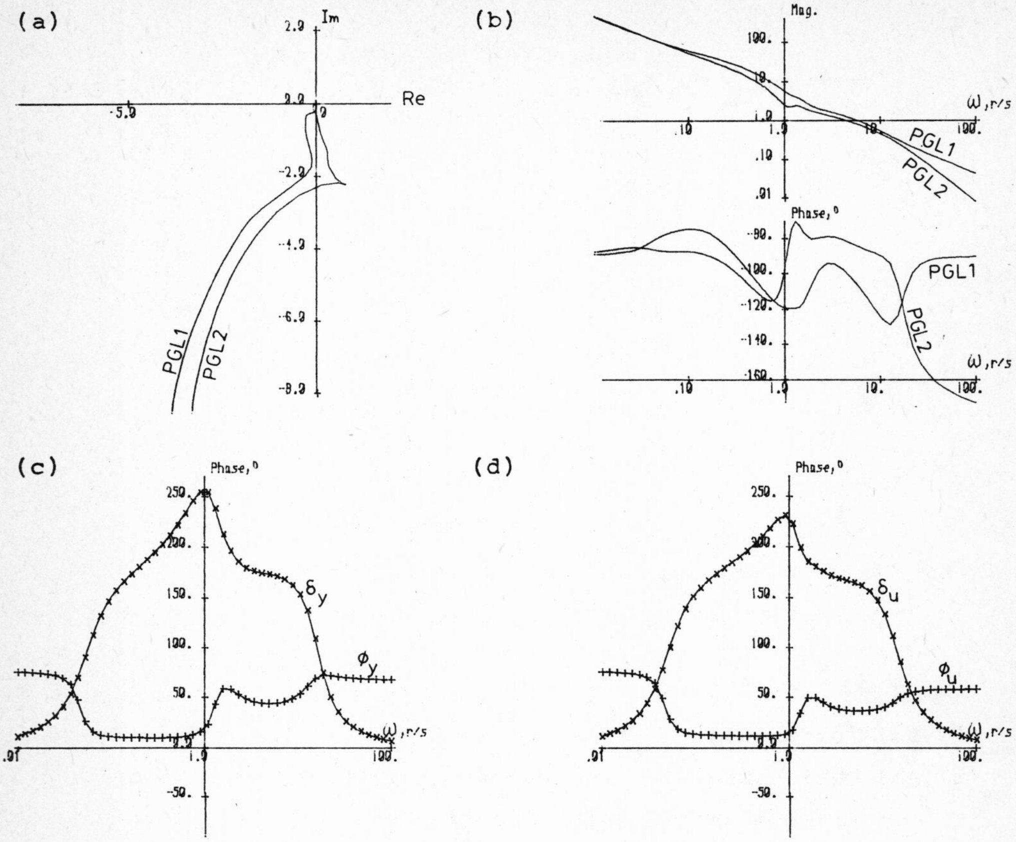

<u>Fig.5.5</u>

(a),(b) PG loci of the precompensated system G(s)K(s).

(c),(d) Output and input frame angles.

CHAPTER 6 CALCULATING A COMPENSATOR BY NONLINEAR LEAST-SQUARES FITTING

In the previous chapter we calculated a precompensator by first specifying its denominator matrix, and we let the least-squares program estimate the numerator matrix. Moreover, because it is not clear how to select a good denominator matrix, we simply took a common denominator with poles faster than the dominant poles of the system. This has the following shortcomings:

(1) A bad choice of precompensator poles will result in poor fitting in the least-squares step.

(2) The use of a common denominator is unnecessarily restrictive.

For these reasons we wish to explore the degrees of freedom we have available with the denominator matrix. A natural question to ask is whether we can estimate the denominator coefficients by a minimization algorithm, as we did for the numerator. The difficulty is, of course, that in the left MFD, $K(s) = D(s)^{-1}N(s)$, $K(s)$ depends linearly on the coefficients of $N(s)$ but nonlinearly on the coefficients of $D(s)$. Certainly it is always possible to use one of the many existing nonlinear optimization algorithms to search for a set of coefficients of both $N(s)$ and $D(s)$ by minimizing the objective function (the residual sum of squares) defined in (5.1.7). This is, however, bound to be inefficient because of the large number of coefficients involved and the amount of calculations needed to evaluate the objective function at a single point of the coefficient space. Fortunately, there are special features of our problem which enable a more efficient algorithm to be devised. This chapter is devoted to the exploitation of such special features, resulting in a design procedure which simplifies, if not totally removes, step (3) of Algorithm 5.1.1.

The nonlinear fitting algorithm, though more powerful than the linear one, has one additional problem. The compensator, whose poles are now determined by a fitting algorithm, is not guaranteed to be stable unless specific constraints are introduced to ensure this. We shall not include such constraints in the discussion to follow.

§6.1 Problem Formulation

We shall continue to use a left MFD for the precompensator

$$K(s) = D(s)^{-1}N(s) \tag{6.1.1}$$

In general, a proper rational function matrix can have many left MFD's. If we want to estimate the coefficients of both D(s) and N(s), we should remove such redundancies from the description (6.1.1) of K(s). This simply means choosing a canonical form out of the equivalence class of all left MFD's giving rise to the same K(s). In order to economize the number of free parameters and hence computations, we shall employ the Hermite canonical form. Specifically, we assume that, in the left MFD (6.1.1),

(1) D(s) is row reduced and that

$$d_i := \deg\left(\mathrm{row}_i(D(s))\right) \geq \deg\left(\mathrm{row}_i(N(s))\right)$$

so that K(s) is proper (see §5.3);

(2) D(s) is in column-Hermite canonical form (e.g. see [KAI, pp.375 & 476]), that is D(s) is upper triangular, with each diagonal element monic and of higher degree than any other element in the same column (if a diagonal element is 1, then all other elements in that column are 0).

Let

D =

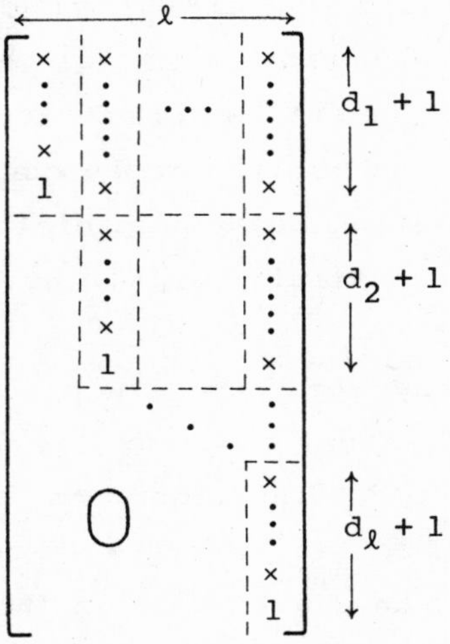

(6.1.2)

be made up from the coefficients of the polynomial entries of $D(s)$ so that

$$D(s) = S(s)D \tag{6.1.3}$$

where $S(s)$ is as defined in (5.3.1). In (6.1.2), the ×'s represent possibly nonzero coefficients. Note that

(1) the lower triangular block is zero because $D(s)$ is upper triangular.

(2) since diagonal elements of $D(s)$ are monic, so the trailing coefficient of each diagonal column-block is 1.

(3) in the (i,j)th column-block $(i<j)$, only the first $\min(d_i, d_j-1)$ coefficients are possibly nonzero. This is because

$$D(s) \text{ is row reduced} \Rightarrow \deg(d_{ij}(s)) \leq \deg(d_{ii}(s))$$
$$D(s) \text{ is column-Hermite} \Rightarrow \deg(d_{ij}(s)) < \deg(d_{jj}(s))$$

Again, as in (5.3.2), we write $N(s)$ as

$$N(s) = S(s)N \tag{6.1.4}$$

Now we modify Algorithm 5.1.1 to

<u>Algorithm 6.1.1 (RFNDP: Fitting a Left MFD Precompensator)</u>

(1) Choose a frequency list $\{\omega_1, \ldots, \omega_\rho, \ldots, \omega_n\}$. For each frequency ω_ρ, do an SVD of the given system $G(s) \in \mathbb{R}(s)^{m\times \ell}$:

$$G_\rho := G(j\omega_\rho) = Y_\rho \Sigma_{G\rho} U_\rho^* \quad \in \mathbb{C}^{m\times \ell}$$

(2) Specify $\Gamma_Q(s) = \mathrm{diag}(\gamma_{Qi}(s))_{i=1}^{r}$, the characteristic gain loci for the desired precompensated system, where $r := \min(m,\ell)$. At each ω_ρ, evaluate

$$Q_\rho := Y_\rho \Gamma_Q(j\omega_\rho) Y_\rho^*$$

(3) Specify the row degrees d_i $(i=1,..,\ell)$ of the denominator matrix of

$$K(s) = D(s)^{-1}N(s) = [S(s)D]^{-1}[S(s)N] \quad \in \mathbb{R}(s)^{\ell\times m}$$

Choose a $D_o(s)$ to be used as a starting point in the minimization step (5) below.

(4) Choose a weighting matrix $W(s) \in \mathbb{R}(s)^{m \times m}$ and let $W_\rho := W(j\omega_\rho)$.

(5) Define the error matrix $E_\rho(D,N)$ to be

$$E_\rho(D,N) := G_\rho [S(j\omega_\rho)D]^{-1}[S(j\omega_\rho)N] - Q_\rho \qquad \rho = 1,\ldots,n \qquad (6.1.5)$$

and determine the coefficient matrices D, N by solving the problem:

$$\underset{D,N}{\text{minimize}} \sum_{\rho=1}^{n} \| E_\rho(D,N) \|_{W_\rho}^2 \qquad (6.1.6)$$

End of Algorithm 6.1.1

Note that in (6.1.5), we have indicated explicitly, for clarity, that the error of fitting depends on D and N. In order not to be too involved in details which will be rather technical, we shall simply take $W(s) = \mathbb{1}_m$ (i.e. no weighting), but we remark that there is no difficulty in adding on a weighting.

It remains to solve the problem formulated in step (5) of Algorithm 6.1.1, which is a nonlinear least-squares fitting problem. If we rewrite (6.1.5) as

$$E_\rho(D,N) = \Phi_\rho(D)N - Q_\rho \qquad \rho = 1,\ldots,n \qquad (6.1.7)$$

where $\Phi_\rho(D) := G_\rho [S(j\omega_\rho)D]^{-1} S(j\omega_\rho)$

then it is clear that E(D,N) depends linearly on N and nonlinearly on D.

Let

$$E(D,N) := \begin{bmatrix} \operatorname{Re} E_1(D,N) \\ \operatorname{Im} E_1(D,N) \\ \hline \cdots \\ \hline \operatorname{Re} E_n(D,N) \\ \operatorname{Im} E_n(D,N) \end{bmatrix} \quad \Phi(D) := \begin{bmatrix} \operatorname{Re}\Phi_1(D) \\ \operatorname{Im}\Phi_1(D) \\ \hline \cdots \\ \hline \operatorname{Re}\Phi_n(D) \\ \operatorname{Im}\Phi_n(D) \end{bmatrix} \quad Q := \begin{bmatrix} \operatorname{Re} Q_1 \\ \operatorname{Im} Q_1 \\ \hline \cdots \\ \hline \operatorname{Re} Q_n \\ \operatorname{Im} Q_n \end{bmatrix} \qquad (6.1.8)$$

Then the n equations of (6.1.7) can be stacked together to give a single real matrix equation:

$$E(D,N) = \Phi(D)N - Q$$

and (6.1.6) becomes

$$\underset{D,N}{\text{minimize}} \; \|E(D,N)\|^2 \;\equiv\; \underset{D,N}{\text{minimize}} \; \|\Phi(D)N - Q\|^2 \qquad (6.1.9)$$

We have now put our problem in a form which has appeared in the numerical analysis literature. We shall recall some relevant results and then outline a procedure for solving (6.1.9) in the next section.

§6.2 A Least-Squares Problem whose Variables Separate

The least-squares problem (6.1.9), in which the parameters to be estimated separate into linear and nonlinear sets has been studied in [GOL],[KAU],[KRO] and efficient algorithms, which take advantage of the special features of the problem, are available. The algorithms are based on the observation that the linear parameters (in N) and the nonlinear parameters (in D) can be solved in two independent steps.

Notice that in (6.1.9), if D is determined, then the N which minimizes $\|\Phi(D)N - Q\|^2$ is given by (see §5.2)

$$N = \Phi(D)^{\dagger} Q \qquad (6.3.1)$$

where we have assumed that $\Phi(D)$ is full rank. Hence we can rewrite (6.1.9) as

$$\begin{aligned}
&\underset{D,N}{\text{minimize}} \; \|\Phi(D)N - Q\|^2 \\
&= \min_{D}\Big(\min_{N} \|\Phi(D)N - Q\|^2 \Big) \\
&= \min_{D} \|\Phi(D)\Phi(D)^{\dagger} Q - Q\|^2 \qquad \text{by (6.3.1)} \\
&= \min_{D} \|P_{\Phi(D)} Q - Q\|^2 \\
&= \min_{D} \|P^{\perp}_{\Phi(D)} Q\|^2 \qquad (6.3.2)
\end{aligned}$$

where $P_{\Phi(D)} := \Phi(D)\Phi(D)^{\dagger}$

and $P^{\perp}_{\Phi(D)} := I - P_{\Phi(D)}$

are the orthogonal projectors onto $\mathscr{R}(\Phi(D))$ and its orthogonal complement. We see that (6.3.2) is now independent of N. This suggests the following strategy for solving (6.1.9):

(VS1) First solve (6.3.2) for D by a nonlinear least-squares technique.

(VS2) Suppose a minimizing $\hat{D}$ has been found in step (VS1), we set

$$\hat{N} = \Phi(\hat{D})^{\dagger} Q \tag{6.3.3}$$

Then the pair $(\hat{D}, \hat{N})$ minimizes $\|\Phi(D)N - Q\|^2$.

A rigorous statement and proof of this can be found in [GOL, Theorem 2.1]. As far as step (VS1) is concerned, we shall use the generalized Gauss-Newton iteration, as follows.

Define

$$E(D) := P^{\perp}_{\Phi(D)} Q \tag{6.3.4}$$

Now pack both sides of (6.3.4) into column vectors by using the usual technique of introducing the Kronecker product (e.g. see [MAR, pp.8-9])

$$\psi(E(D)) = \left(I_m \otimes P^{\perp}_{\Phi(D)}\right) \psi(Q)$$

where $I_m \otimes P^{\perp}_{\Phi(D)}$:= Kronecker product of I_m and $P^{\perp}_{\Phi(D)}$

and $\psi(X)$:= the column vector obtained by stacking the columns of the matrix X, from left to right, into a single column.

(6.3.2) now becomes

$$\min_D \|E(D)\|^2 = \min_D \|\psi(E(D))\|^2$$
$$= \min_D \|\left(I_m \otimes P^{\perp}_{\Phi(D)}\right)\psi(Q)\|^2 \tag{6.3.5}$$

Applying the generalized Gauss-Newton iteration (e.g. see [ORT, §8.5] or [FLE, Chapter 6]) with step control to the minimization problem (6.3.5) now gives the iterative equation

$$\begin{aligned}\psi(D_{i+1}) &= \psi(D_i) - t_i[\partial(I_m \otimes P^{\perp}_{\Phi(D_i)})\psi(Q)]^{\dagger}\psi(E(D_i)) \\ &= \psi(D_i) - t_i[(I_m \otimes \partial P^{\perp}_{\Phi(D_i)})\psi(Q)]^{\dagger}\psi(E(D_i)) \qquad (6.3.6)\end{aligned}$$

where D_i := the denominator coefficient matrix (see (6.1.3)) at step i (recall that D_o is chosen in step (3) of Algorithm 6.1.1);

∂ denotes the Fréchet derivative (e.g. see [ORT, §3.1] or [GLA, Chapter V §4]) w.r.t. nontrivial elements of D;

and t_i (usually = 1) defines the length of the step from $\psi(D_i)$ and is adjusted to ensure that $\|E(D_{i+1})\| < \|E(D_i)\|$.

Using (6.3.6) as an iterative equation, the main bulk of computations will be the evaluation of the derivative $\partial P^{\perp}_{\Phi(D_i)}$. This is discussed in detail in [GOL], which relates $\partial P^{\perp}_{\Phi(D_i)}$ to the more readily available $\partial\Phi(D_i)$. With some changes in notation and dropping the subscript i, we quote (see [GOL, equations (4.2), (4.5) & (5.4)]):

$$\partial P^{\perp}_{\Phi(D)} = -(P^{\perp}_{\Phi(D)}\partial\Phi(D))\Phi(D)^{\dagger} - \Phi(D)^{\dagger}(P^{\perp}_{\Phi(D)}\partial\Phi(D))^{T} \qquad (6.3.7)$$

In (6.3.7), the derivative $\partial\Phi(D)$ is given by (see (6.1.8))

$$\partial\Phi(D) = \begin{bmatrix} \cdots \\ \hline \mathrm{Re}\,\partial\Phi_{\rho}(D) \\ \mathrm{Im}\,\partial\Phi_{\rho}(D) \\ \hline \cdots \end{bmatrix} \qquad (6.3.8a)$$

Using a formula for the derivative of the inverse of an operator (e.g. see [GLA, Chapter V, problem 134])

$$\begin{aligned}\partial\Phi_{\rho}(D) &= \partial[G_{\rho}D(j\omega_{\rho})^{-1}S(j\omega_{\rho})] \\ &= G_{\rho}\,\partial[D(j\omega_{\rho})^{-1}]\,S(j\omega_{\rho})\end{aligned}$$

$$= G_\rho D(j\omega_\rho)^{-1}[\partial D(j\omega_\rho)]D(j\omega_\rho)^{-1}S(j\omega_\rho)$$

$$= G_\rho D(j\omega_\rho)^{-1}S(j\omega_\rho)[\partial D]D(j\omega_\rho)^{-1}S(j\omega_\rho) \qquad (6.3.8b)$$

(6.3.7) and (6.3.8a,b) now provide an explicit means for calculating $\partial P^{\perp}_{\Phi(D)}$. These, together with (6.3.6) constitute a set of equations for an iterative procedure for solving (6.3.2). Note that in the above set of equations, we have regarded $\partial P^{\perp}_{\Phi(D)}$, $\partial\Phi(D)$, etc. as tridimensional tensors, in a natural way. For example, $\partial\Phi(D)$ is formed from ν (= number of free parameters in D) slabs of matrices, each of which is the partial derivative of $\Phi(D)$ w.r.t. one of the parameters of D. The tridimensional tensor $\left(I_m \otimes \partial P^{\perp}_{\Phi(D_i)}\right)$, when multiplied with the vector $\psi(Q)$ (see (6.3.6)) will be contracted to a two dimensional matrix. By virtue of the requirement of matching dimensions in the multiplications, it should be clear how the vectors, matrices and tensors in (6.3.6), (6.3.7) and (6.3.8a,b) interact.

Having completed the step (VS1), the next step (VS2) is then a direct computation using the result of step (VS1).

This completes our discussion of a solution to (6.1.9) and also the implementation of step (5) of Algorithm 6.1.1. We do not propose to go into numerical or coding details here.

§6.3 Example

We now illustrate Algorithm 6.1.1 by repeating the design carried out in Example 5.4.1 for the plant AUTM.

Example 6.3.1

In order to compare the result we shall obtain in the present design with that of Example 5.4.1, we shall make the same choices as those taken previously. Specifically, in step (2), we specify $\gamma_{Q1}(s)$, $\gamma_{Q2}(s)$ as in Example 4.7.1. In step (3), we choose $D_o(s) = s(s+2)I_2$ as a starting point for the nonlinear least-squares iteration and in step (4), we use (5.4.1-2) as the weighting matrix. The minimization procedure then yields, after 6 iterations, the precompensator $K(s) = D(s)^{-1}N(s)$ where

$$D(s) = \begin{bmatrix} s(s+0.267) & 0.812s \\ 0 & s(s+1.94) \end{bmatrix}$$

$$N(s) = \begin{bmatrix} 1.48s^2 + 2.33s + 0.230 & -0.428s^2 + 1.07s + 0.251 \\ -0.189s^2 + 0.884s - 1.48 & 0.533s^2 + 5.03s + 1.20 \end{bmatrix}$$

with a weighted relative error of fitting = 0.0880

To guarantee that K(s) has two integrators (see Example 4.7.1), we have constrained the constant terms of the polynomial elements of D(s) to be zero. This means that D(s) has only 3 free parameters, whereas N(s) contains 12 free parameters.

Analysis of the precompensated system gives the graphs of Fig.6.1 to Fig.6.3. They compare favourably with the corresponding results presented in Fig.5.3 and Fig.5.5. However, the improvement is not substantial, as is reflected by the fact that the relative error improves only from 0.139 to 0.0880. This is because in Example 5.4.1, the precompensator poles are already quite well-chosen.

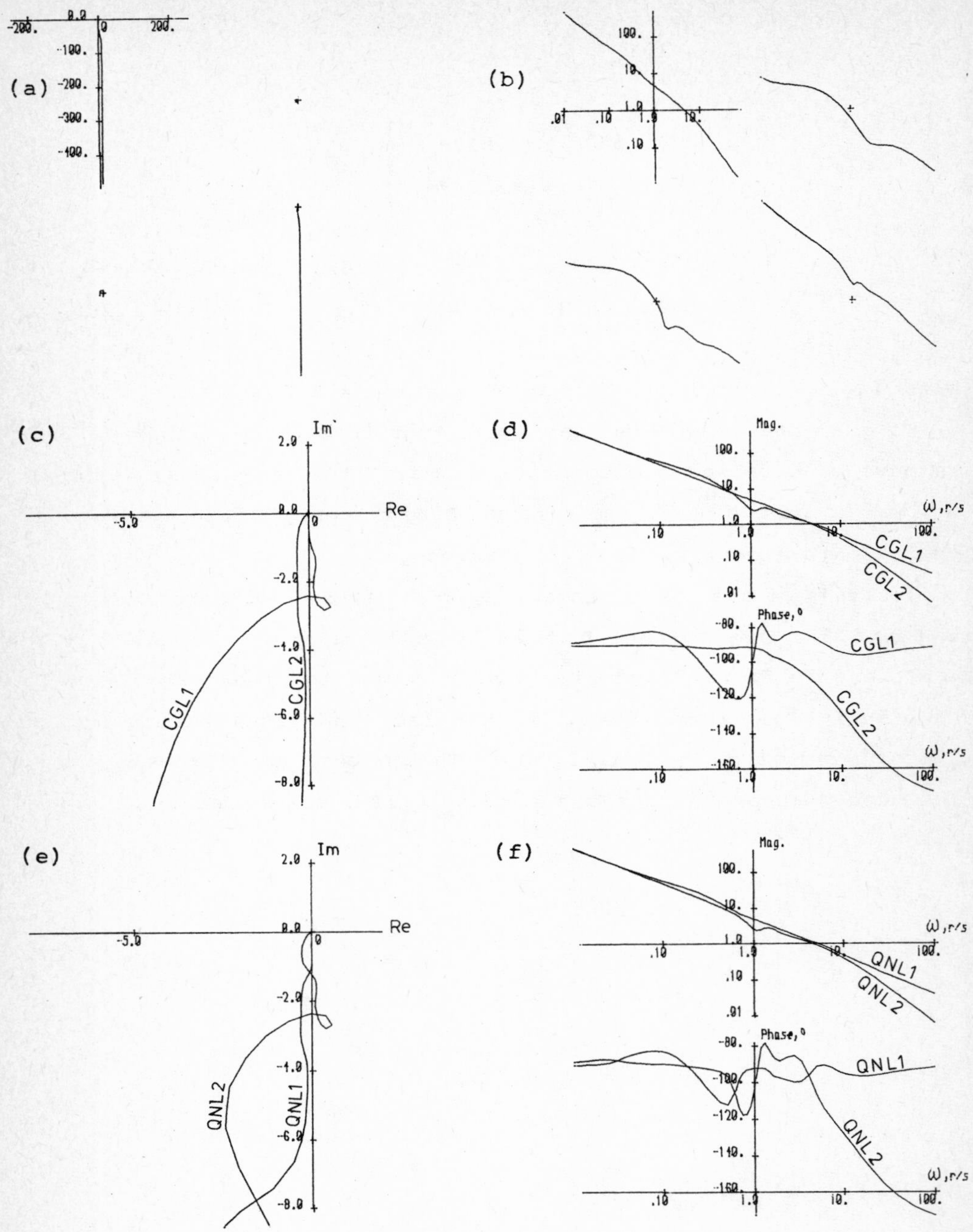

Fig.6.1 After the nonlinear least-squares fitting, results for the precompensated system G(s)K(s) are given above.

(a),(b) Nyquist and Bode magnitude arrays.

(c),(d) Characteristic gain loci.

(e),(f) QN loci.

(g)

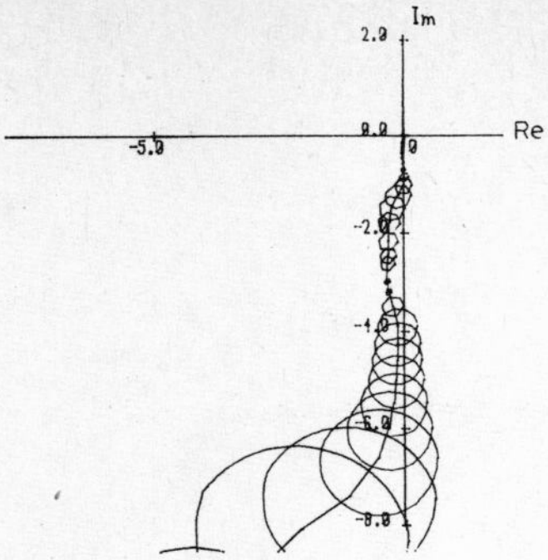

(h)

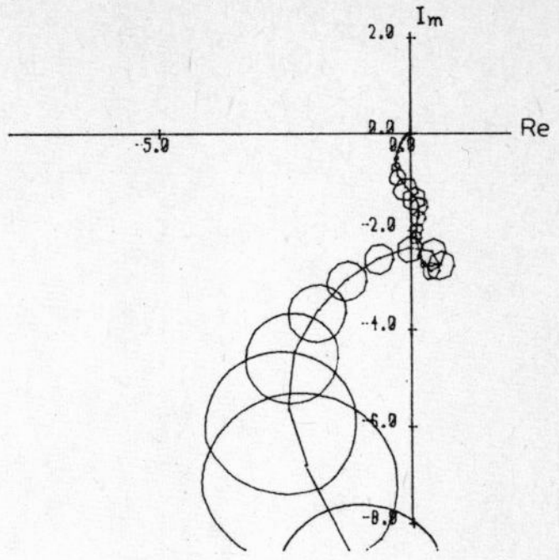

(i)

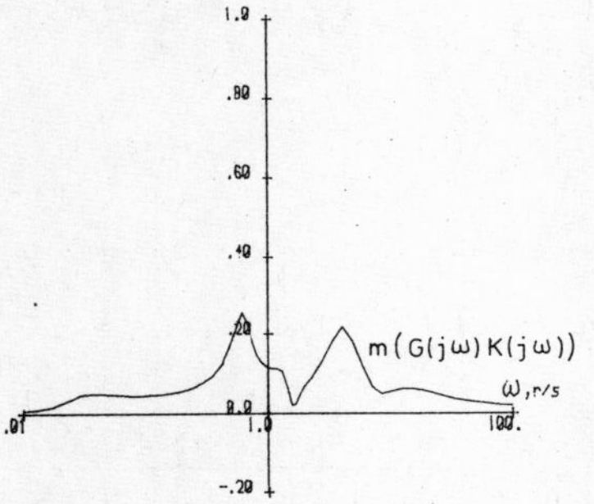

<u>Fig.6.1</u> (continued)

(g),(h) QNLi taken with the bands $B_i(j\omega)$ (i=1,2).

(i) Frame misalignment.

(a)

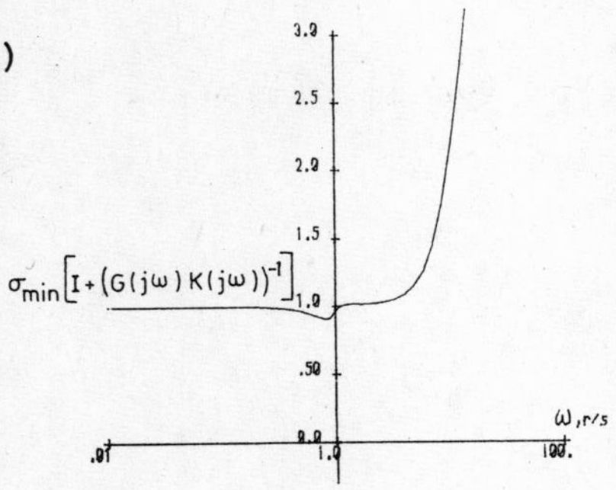

(b)

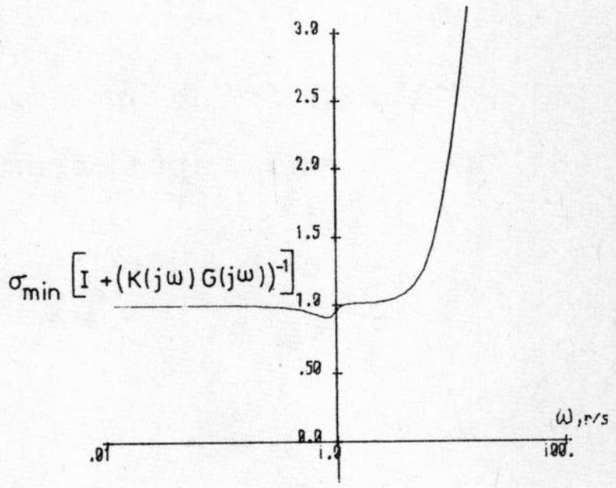

<u>Fig.6.2</u> Robustness measures at (a) plant output; and (b) plant input.

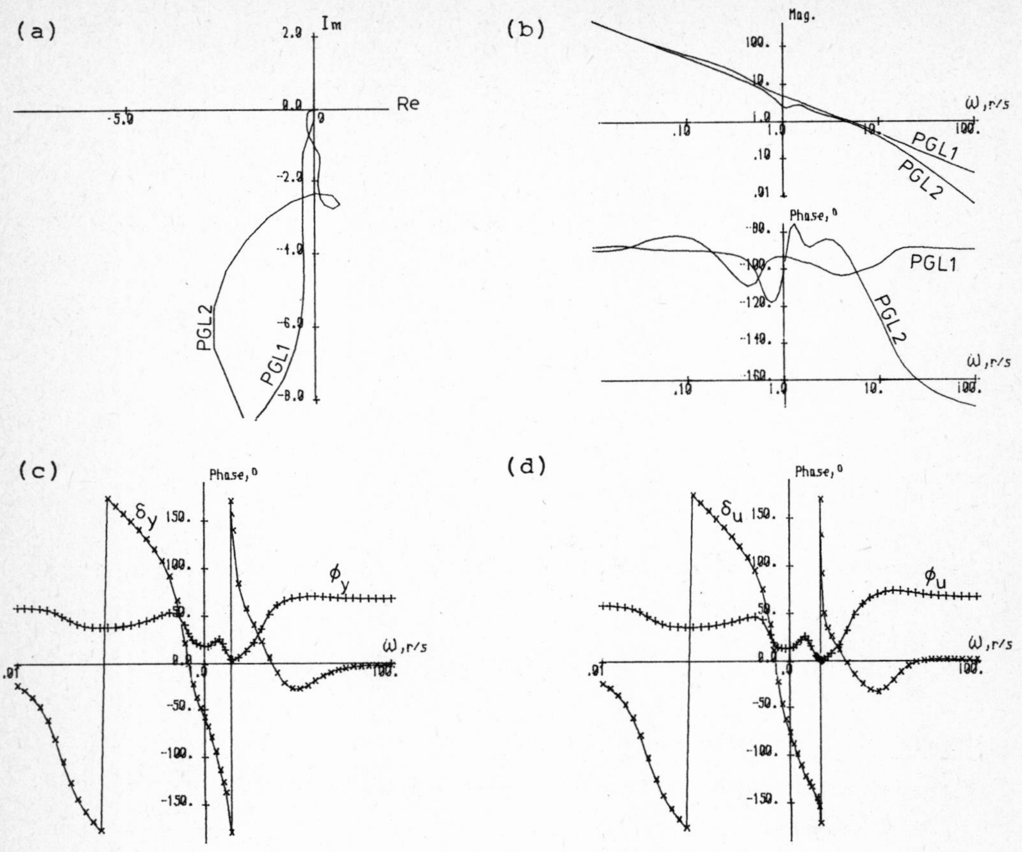

Fig.6.3

(a),(b) PG loci of the precompensated system G(s)K(s).

(c),(d) Output and input frame angles.

CHAPTER 7 EXAMPLES OF THE DESIGN TECHNIQUES

We now illustrate the various concepts and techniques developed in the preceding chapters by applying them to a number of plant models. The first plant has the same number of input and output variables. Subsequent plants have different numbers of inputs and outputs; a discussion of the "non-square plant" case is given before these later examples.

§7.1 A Design Example for a Turbo-Generator

In this section we consider a model of a turbo-generator whose details are given in Appendix E. This system is open-loop stable. Although there is a pair of right-half-plane zeros, their locations are so far beyond the bandwidth of the system that their effect can be neglected. The Nyquist and Bode magnitude arrays (see Appendix E, Fig.E.1) show that the elements of the second column are of more than two orders of magnitude larger than those of the first column. This is simply a scaling problem. We can re-scale the system and orthogonalize the dc eigenframe simultaneously, using a constant precompensator, as follows.

First, we do an SVD of G(0):

$$
\begin{aligned}
G(0) &= Y_o \Gamma_o U_o^T \\
&= \begin{bmatrix} .662 & .750 \\ -.750 & .662 \end{bmatrix} \begin{bmatrix} 903 & 0 \\ 0 & .657 \end{bmatrix} \begin{bmatrix} -.00134 & -1.00 \\ -1.00 & .00134 \end{bmatrix}^T
\end{aligned}
$$

To reduce the larger gain and multiply up the smaller gain, each by a factor of 10, we define (see §3.6.1)

$$
\begin{aligned}
K_o &:= K_o(0.1, 10) \\
&= U_o \operatorname{diag}(0.1, 10)\, Y_o^T
\end{aligned}
$$

$$= \begin{bmatrix} -7.50 & -6.62 \\ -.0561 & .0838 \end{bmatrix}$$

The results of this preliminary precompensation are given in Fig.7.1, which shows that $G(s)K_o$ will actually be closed-loop unstable. The outstanding feature of the system is clearly a pair of lightly damped resonant poles at $-0.351 \pm j6.34$, which is typical of power generator systems. This, together with the higher than first order infinite zeros, are likely causes of difficulties of feedback control. Furthermore, the linear model, obtained by a linearization at a nominal operation point, is valid only when the input signals are kept below some saturation level. Hence excessive controller gains are not realistic from a practical point of view.

Bearing all these in mind, we proceed to do a feedback controller design using Algorithm 6.1.1. After a number of trials, we finally specify the desired characteristic gains $\gamma_{Q1}(s)$, $\gamma_{Q2}(s)$ and the weighting matrix $W(s)$ to be

$$\gamma_{Q1}(s) = \frac{500(s+0.5)}{s(s+5)(s+20)}$$

$$\gamma_{Q2}(s) = \frac{25000(s+0.5)}{s(s+5)(s+20)(s+50)}$$

$$W(s) = |w(s)|^2 \, \mathbb{1}_2$$

where $$w(s) = \frac{100s+1}{0.1s+1}$$

It is readily checked that the desired characteristic gains are consistent with the ideas of Chapter 4. Graphs for $\gamma_{Q1}(j\omega)$, $\gamma_{Q2}(j\omega)$, the desired response and the weighting function $|w(j\omega)|$ are shown in Fig.7.2.

We then go on to do the nonlinear-least squares fitting, which yields the precompensator

$$K(s) = D(s)^{-1}N(s)$$

where

$$D(s) = \begin{bmatrix} s^3 + 61.6s^2 + 1670s & 0 \\ 0 & s^3 + 114s^2 + 1860s \end{bmatrix}$$

$$N(s) = \begin{bmatrix} 41.3s^3 + 66.5s^2 + 1580s + 284 & -9.78s^3 - 136s^2 + 657s + 288 \\ 44.2s^3 + 44.5s^2 + 1510s + 230 & -5.77s^3 + 408s^2 + 125s + 331 \end{bmatrix}$$

with a weighted relative error of fitting = 0.277

We should point out that :

(1) A 6-state controller is used because a lower order controller does not give satisfactory fitting.

(2) The denominator is diagonal (rather than Hermite) because we have constrained it to be so. The reason being that, for this example, allowing the (1,2) element of D(s) to be nonzero results in a D(s) with a large off-diagonal term and in a N(s) with large coefficients. Besides, the improvement is not substantial over the precompensator given above.

The precompensator K(s) has poles at

$$\{ 0, 0, -19.9, -30.8 \pm j26.8, -93.7 \}$$

and zeros at

$$\{ -0.152 \pm j0.0942, -0.534 \pm j6.16, -1.01, -116 \}$$

Graphs for $G(s)K_0K(s)$ are given in Fig.7.3. They show that the system's resonant modes are quite well damped out by the controller, which has a pair of zeros approximately cancelling the system's pair of resonant poles. This is of course what we have implicitly asked for via the specified desired response.

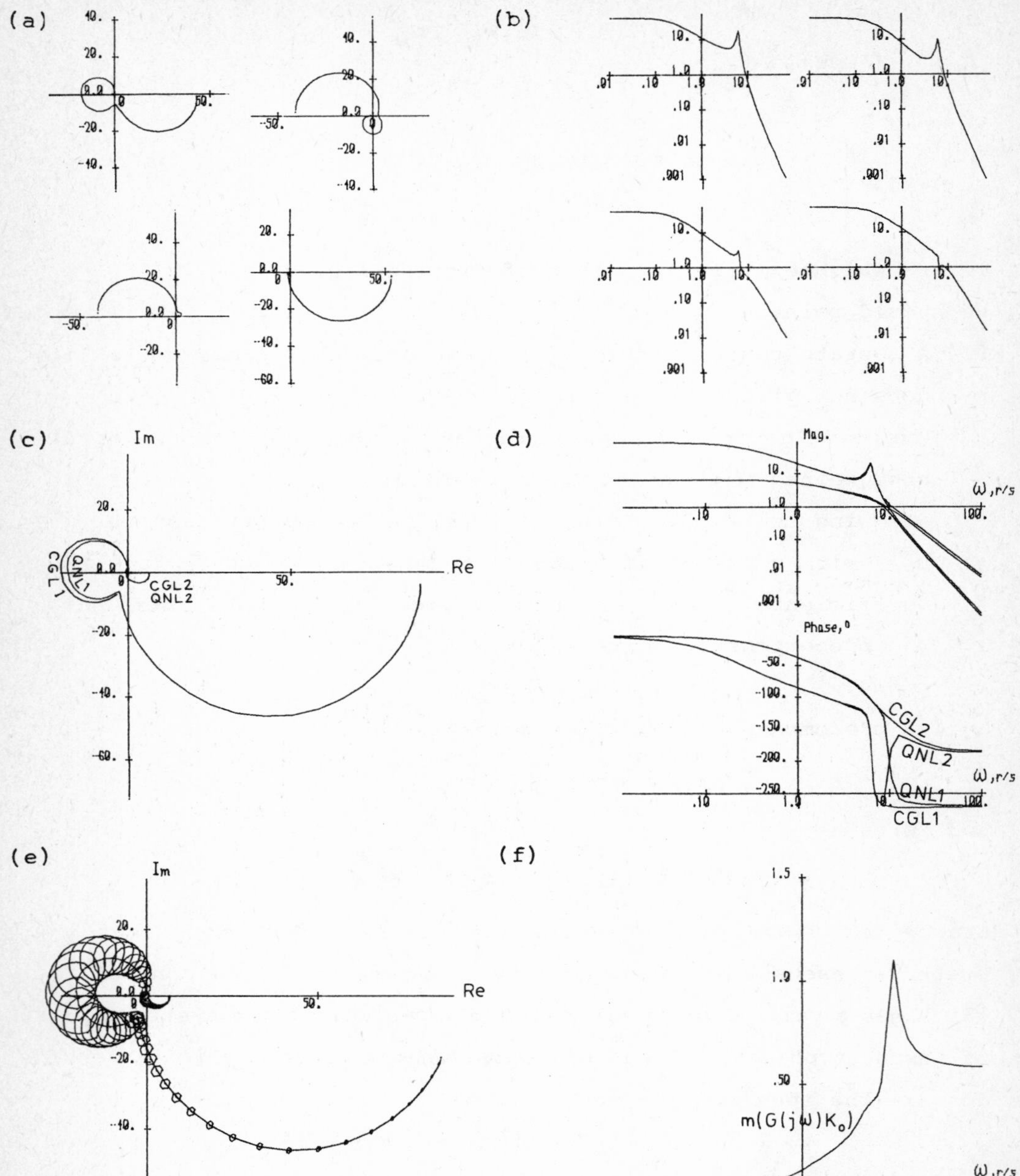

Fig.7.1 A preliminary constant precompensator gives the system $G(s)K_o$.

(a),(b) Nyquist and Bode magnitude arrays.

(c),(d) Characteristic gain loci and QN loci.

(e) QN loci with bands.

(f) Frame misalignment.

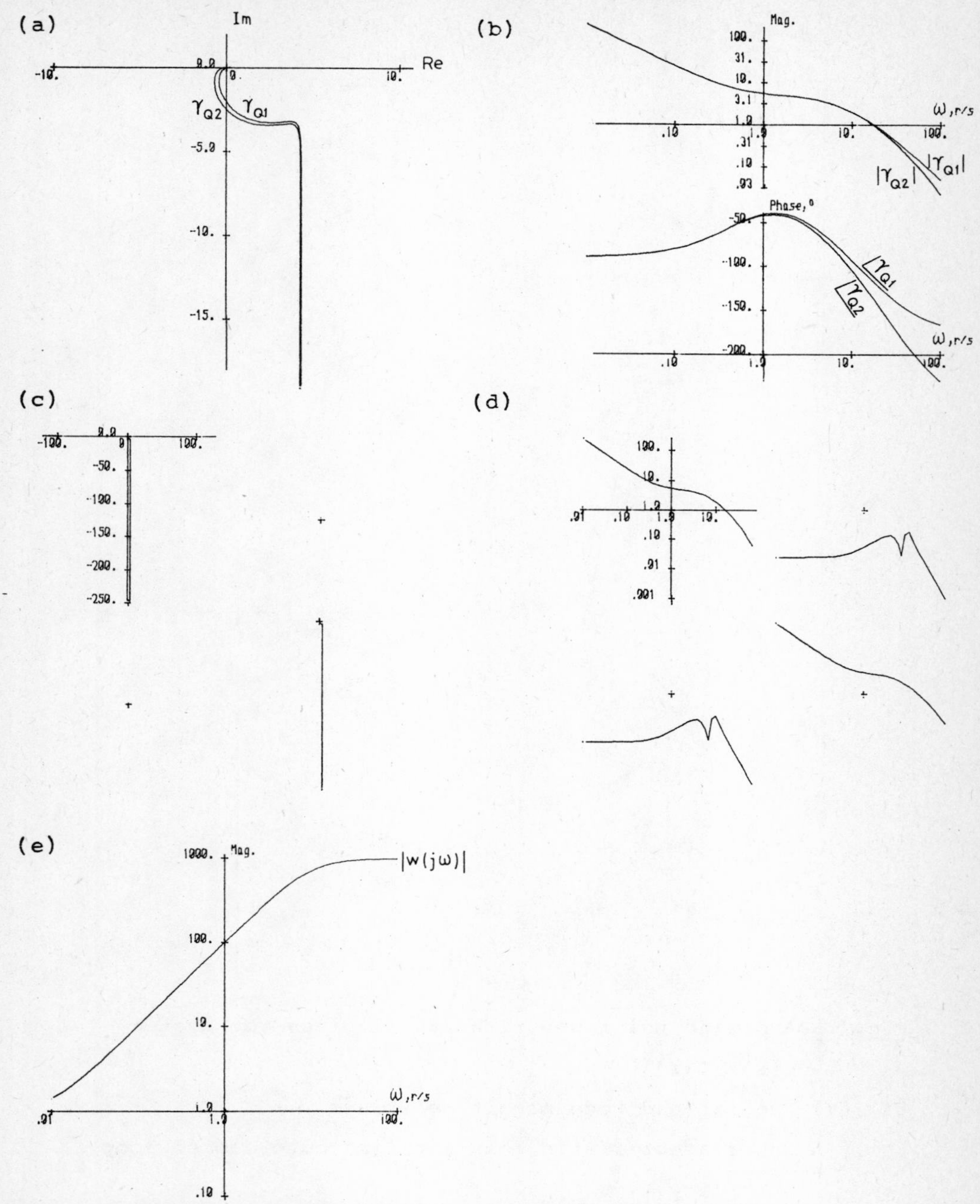

Fig.7.2 Specifications for the nonlinear least-squares fitting.

(a),(b) The desired characteristic gains $\gamma_{Q1}(j\omega)$, $\gamma_{Q2}(j\omega)$.

(c),(d) Nyquist and Bode magnitude arrays of the desired response.

(e) The weighting function $|w(s)|$.

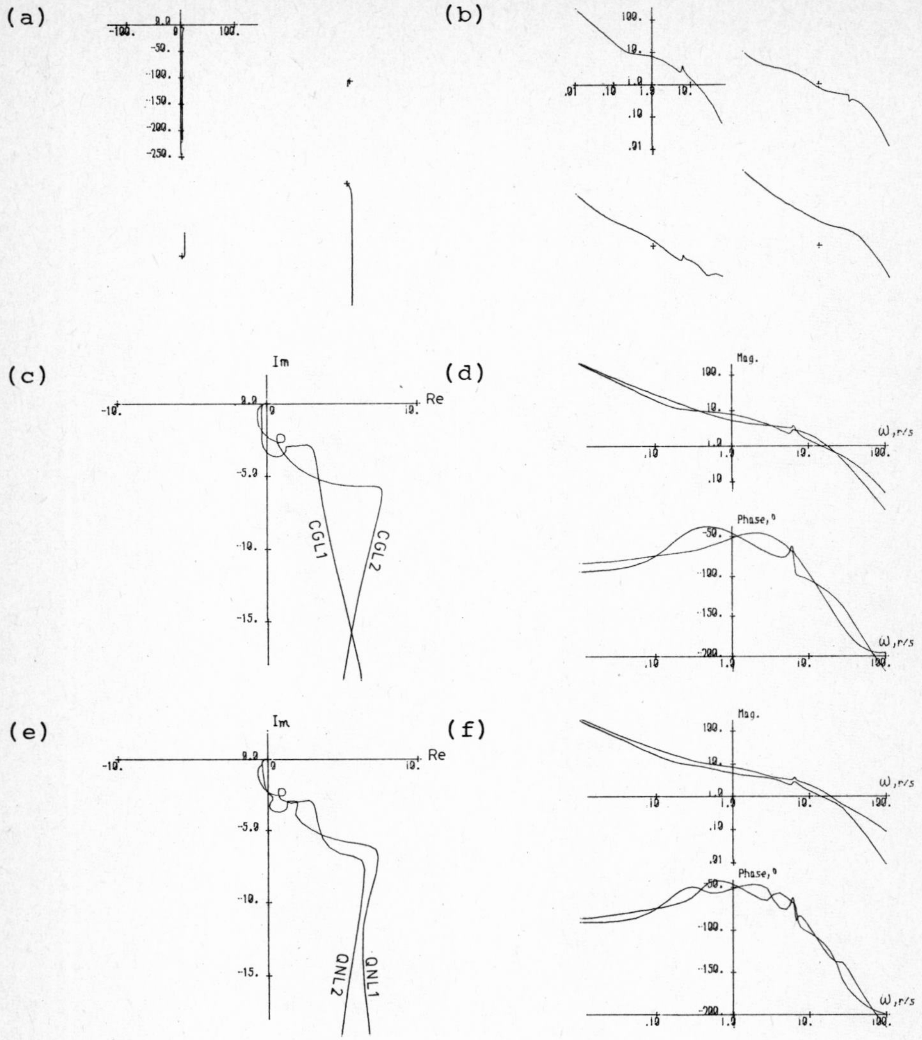

Fig.7.3 A design using Algorithm 6.1.1 gives the system $G(s)K_o K(s)$.

(a),(b) Nyquist and Bode magnitude arrays.

(c),(d) The characteristic gain loci indicate closed-loop stability.

(e),(f) The QN loci show that the resonant modes are well damped.

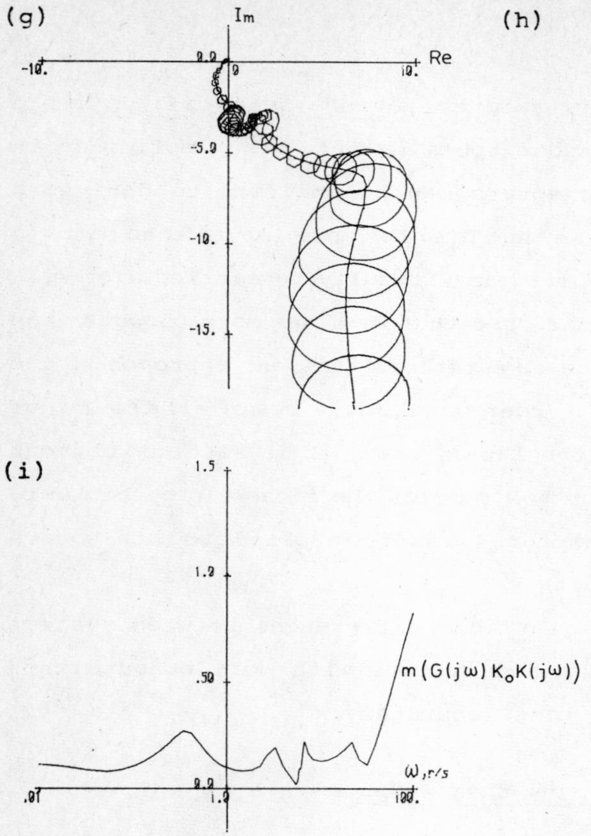

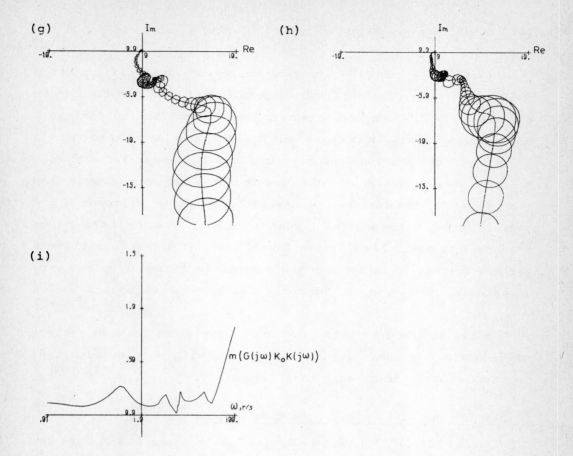

Fig.7.3 (continued)

(g),(h) QNL1 and QNL2 with bands.

(i) Frame misalignment.

§7.2 Non-Square Systems

A system $G(s) \in \mathbb{R}(s)^{m\times\ell}$ is said to be <u>non-square</u> if it has a different number of inputs and outputs ($m \neq \ell$). The optimization algorithms of the previous chapters can be modified to deal with non-square systems. In the case that G(s) has more inputs than outputs ($m < \ell$), all the previous results can virtually be carried over with no extra work. However, in the case that G(s) has more outputs than inputs ($m > \ell$), the designer needs a rather different approach to the overall compensation scheme in order to make full use of all the output measurements available. In the latter case, it is also not obvious without further investigation how some of the ideas (in particular, robustness) developed in Chapter 4 can be applied to non-square systems.

Because of the intrinsic practical differences between systems with more inputs than outputs and systems with more outputs than inputs, we shall treat each case separately.

§7.2.1 Systems with More Inputs than Outputs

Let $G(s) \in \mathbb{R}(s)^{m\times\ell}$ have more inputs than outputs ($m < \ell$), and let $K(s) \in \mathbb{R}(s)^{\ell\times m}$ be a precompensator for G(s). The usual negative-unity-feedback configuration gives the closed-loop system of Fig.7.4.

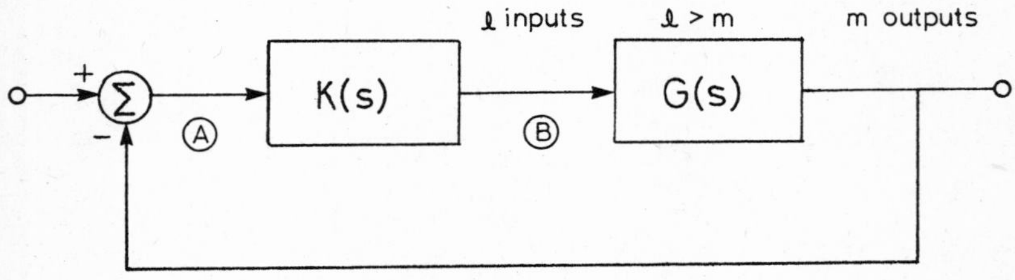

Fig.7.4

With respect to the loop-breaking points Ⓐ or Ⓑ, the return-ratio matrices are $G(s)K(s) \in \mathbb{R}(s)^{m\times m}$ and $K(s)G(s) \in \mathbb{R}(s)^{\ell\times\ell}$.

For $s = j\omega_\rho \in D_{NYQ}$, let $G(j\omega_\rho)$ have an SVD:

$$G(j\omega_\rho) = Y_\rho \Sigma_{G\rho} U_\rho^* \tag{7.2.1}$$

where $Y_\rho \in \mathbb{C}^{m\times m}$ is unitary and $U_\rho^* \in \mathbb{C}^{m\times \ell}$ is subunitary. Using the reversed-frame-normalization idea, it would be desirable for $K(j\omega_\rho)$ to have the form

$$K(j\omega_\rho) \approx U_\rho \operatorname{diag}(\cdot) Y_\rho^* \tag{7.2.2}$$

so that

$$G(j\omega_\rho)K(j\omega_\rho) \approx Y_\rho \Gamma_Q(j\omega_\rho) Y_\rho^* \in \mathbb{C}^{m\times m} \tag{7.2.3}$$

$$K(j\omega_\rho)G(j\omega_\rho) \approx U_\rho \Gamma_Q(j\omega_\rho) U_\rho^* \in \mathbb{C}^{\ell\times \ell} \tag{7.2.4}$$

for some specified diagonal $\Gamma_Q(s)$. This means that the return-ratio matrices with respect to both loop-breaking points are approximately normal which would in turn ensure insensitivity of the characteristic gain loci to perturbations (see §1.6). As already described in detail in Chapters 5 and 6, we can search for an appropriate $K(s)$ via an optimization problem defined by choosing a $\Gamma_Q(j\omega)$ and minimizing the difference between the two sides of either (7.2.3) or (7.2.4). It should be noted that requiring $G(j\omega_\rho)K(j\omega_\rho)$ to fit the right hand side of (7.2.3) is not the same as requiring $K(j\omega_\rho)G(j\omega_\rho)$ to fit the right hand side of (7.2.4). In view of the larger dimension of $K(j\omega_\rho)G(j\omega_\rho)$, it is apparent that the latter minimization problem is more stringent. If the fitting error for (7.2.4) (i.e. the difference between the two sides of (7.2.4)) turns out to be small, then $K(j\omega)$ will be given by (7.2.2). (7.2.3), however, does <u>not</u> have similar implications for (7.2.2).

The above discussion indicates that when designing a precompensator for a non-square system with more inputs than outputs, the precompensator should be determined by a minimization problem based on (7.2.4) (rather than one based on (7.2.3)):

$$\underset{K(s)\,\in\,\mathcal{K}}{\text{minimize}} \sum_{\rho=1}^{n} \left\| K(j\omega_\rho)G(j\omega_\rho) - U_\rho \Gamma_Q(j\omega_\rho) U_\rho^* \right\|^2_{W(j\omega_\rho)} \tag{7.2.5}$$

where $\mathcal{K}$ is some specified parameter space for $K(s)$. (7.2.5) can be solved by first transposing to get the equivalent problem

$$\underset{K(s)\,\varepsilon\,\mathcal{K}}{\text{minimize}} \sum_{\rho=1}^{n} \left\| G(j\omega_\rho)^T K(j\omega_\rho)^T - \bar{U}_\rho \Gamma_Q(j\omega_\rho) U_\rho^T \right\|^2_{W(j\omega_\rho)^T}$$

and then making use of either Algorithm 5.1.1 or Algorithm 6.1.1 to determine $K(s)^T$.

§7.2.2 Systems with More Outputs than Inputs

If a system $G(s) \,\varepsilon\, \mathbb{R}(s)^{m\times\ell}$ has more outputs than inputs $(m > \ell)$, then we can expect to have independent control over at most ℓ linear combinations of the m output variables. Possibly after doing an output-coordinate transformation, we may assume without loss of generality that the first ℓ outputs are to be controlled. That is, if we partition the output vector $y := G(s)u$ as

$$y = \begin{bmatrix} c \\ \hline z \end{bmatrix} \begin{matrix} \updownarrow \ell \\ \updownarrow m-\ell \end{matrix} := \begin{bmatrix} G_1(s) \\ \hline G_2(s) \end{bmatrix} u \qquad (7.2.6)$$

where $G_1(s) \,\varepsilon\, \mathbb{R}(s)^{\ell\times\ell}$ is the upper square block of $G(s)$, then the ℓ-dimensional subvector c of y contains the controlled variables and z is an $(m-\ell)$-dimensional vector of extra available measurements. We shall represent (7.2.6) diagrammatically by Fig.7.5.

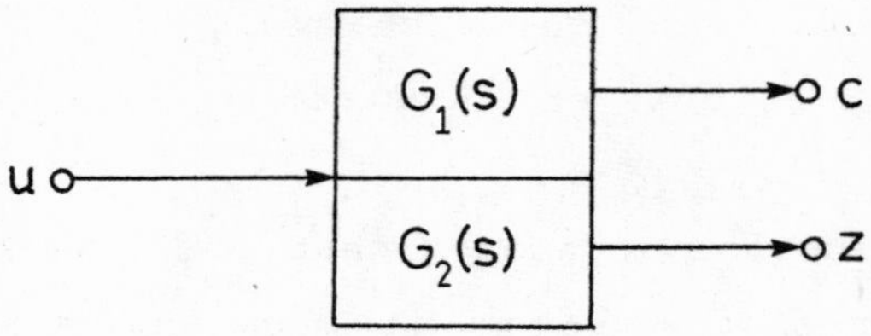

Fig.7.5

Now consider the feedback configuration as shown in Fig.7.6, consisting of an inner feedback loop around $F \,\varepsilon\, \mathbb{R}^{\ell\times m}$, $L \,\varepsilon\, \mathbb{R}^{\ell\times\ell}$ and an outer feedback loop around a dynamic precompensator $K(s) \in \mathbb{R}(s)^{\ell \times \ell}$. The idea behind this configuration is that we manipulate the inner feedback loop in such a way as to use the full set of all available measurements to the best advantage to shape the transmittance from v to c into a suitable form for the dynamic precompensator to inject a suitably high gain round the outer loop. Closing the loop from c to e by a negative-unity-feedback path will then force the controlled variables to track some given reference input r, and to reject disturbances acting on the plant.

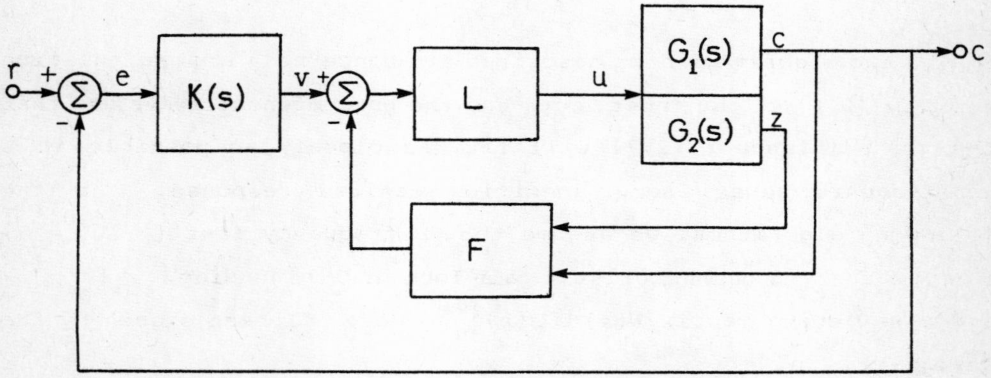

Fig.7.6

We shall now describe two optimization algorithms and a design precedure for determining K(s), L and F in the feedback arrangement of Fig.7.6. Recall that in the case of a square plant, the minimization problem involved is formulated in terms of fitting some prescribed open-loop response. For this configuration, we shall formulate a minimization problem in terms of fitting some desired response from e to c (see Fig.7.6), with the negative-unity-feedback path open while keeping the inner loop closed (see Fig.7.7). The transmittance from e to c is given by

$$T_{ec}(s) := G_1(s)L(I + FG(s)L)^{-1}K(s) \qquad (7.2.7)$$

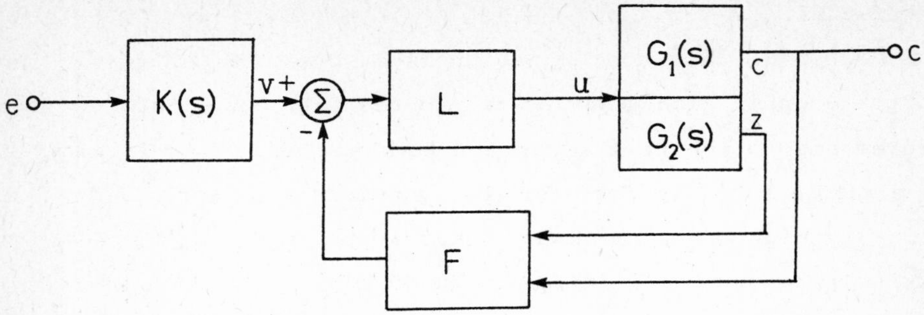

Fig.7.7

The two algorithms next described are concerned with manipulating the pair (L,F) of the inner loop, and the precompensator K(s) so that the transmittance (7.2.7) will fit, as closely as possible in a least-square sense, some specified desired response. In the following algorithms, we assume that a frequency list $\{\omega_1,\ldots,\omega_n\}$, a set of desired characteristic gain loci in $\Gamma_Q(s) := \text{diag}\{\gamma_{Qi}(s)\}_{i=1}^{\ell}$ and a weighting matrix $W(s) \in \mathbb{R}(s)^{\ell\times\ell}$ have all been chosen by the designer.

Algorithm 7.2.1 (Fitting a precompensator K(s) for a fixed (L,F))

(1) Choose a pair of matrices (L,F) for the inner loop.

(2) Define an $\ell\times\ell$ square system H(s) (transmittance from v to c in Fig.7.7) by

$$H(s) := G_1(s)L\left(I + FG(s)L\right)^{-1} \in \mathbb{R}(s)^{\ell\times\ell} \qquad (7.2.8)$$

At each frequency ω_ρ, do an SVD of $G_1(j\omega_\rho)$:

$$G_1(j\omega_\rho) = Y_\rho \Sigma_{G\rho} U_\rho^* \qquad \rho = 1,\ldots,\omega_n$$

(3) Specify a parameter space $\mathcal{K}$, in terms of the numerator and denominator coefficients of a left MFD $D(s)^{-1}N(s)$, for the precompensator K(s). Determine K(s) by solving

$$\underset{K(s)\,\varepsilon\,\mathcal{K}}{\text{minimize}} \sum_{\rho=1}^{n} \left\| H(j\omega_\rho)K(j\omega_\rho) - Y_\rho \Gamma_Q(j\omega_\rho) Y_\rho^* \right\|^2_{W(j\omega_\rho)} \qquad (7.2.9)$$

End of Algorithm 7.2.1

Algorithm 7.2.2 (Adjusting the inner loop pair (L,F) for a fixed K(s))

(1) Choose a precompensator $K(s) \,\varepsilon\, \mathbb{R}(s)^{\ell\times\ell}$.

(2) At each frequency ω_ρ, do an SVD of $G_1(j\omega_\rho)$:

$$G_1(j\omega_\rho) = Y_\rho \Sigma_{G\rho} U_\rho^* \qquad \rho = 1,..,n$$

(3) Specify parameter spaces $\mathcal{L}$, $\mathcal{F}$ for the coefficients of the pair of matrices L, F. Choose some (L_o, F_o) to be used as a starting point in the minimization algorithm. Determine (L,F) by solving

$$\underset{(L,F)\,\varepsilon\,\mathcal{L}\times\mathcal{F}}{\text{minimize}} \sum_{\rho=1}^{n} \left\| G_1(j\omega_\rho)L\left(I+FG(j\omega_\rho)L\right)^{-1}K(j\omega_\rho) - Y_\rho \Gamma_Q(j\omega_\rho) Y_\rho^* \right\|^2_{W(j\omega_\rho)} \qquad (7.2.10)$$

End of Algorithm 7.2.2

The minimization problem (7.2.9) is clearly of the same form as those stated in Algorithms 5.1.1 or 6.1.1 and hence can be solved by the methods given in Chapters 5 or 6. As for the problem (7.2.10), we observe that the sum of squares to be minimized depends nonlinearly on both L and F and hence the best we can do is simply to invoke a nonlinear-least-squares minimization algorithm (e.g. see [FLE, Chapter 6]) to search for such a solution. We shall not go here into further details of implementing such algorithms.

In each of the Algorithms 7.2.1 or 7.2.2, the designer is required to specify either K(s) or (L,F) as the beginning step. The chosen part is then kept fixed while the other is being calculated. The initial choice, however, is often not easy and has a great effect on the success of the subsequent design. It is thus important to provide some possible ways of starting off. A reasonable route is to combine Algorithms 7.2.1 and 7.2.2 into the following design precedure.

Design Procedure 7.2.3

(1) Start the precedure by choosing one of the following options:

(i) Put $K(s) = I$. Go to step 3.

(ii) Put $L = I$, $F = O$. Go to step 2.

(iii) If all the states of the plant being controlled are available, then put $L = I$ and calculate F as the state feedback matrix of a linear-quadratic regulator. Go to step 2.

(2) With the (L,F) determined in the previous step, use Algorithm 7.2.1 to find a K(s). Study the compensated transmittance $T_{ec}(s)$ (see (7.2.7)). In particular, compare $T_{ec}(s)$ with the specified desired response.

(i) If $T_{ec}(s)$ satisfies all design objectives, then stop.

(ii) If $T_{ec}(s)$ falls short of design objectives but has scope for improvement by exploiting further degrees of freedom in adjusting (L,F), then go on to step 3.

(iii) If $T_{ec}(s)$ is unsatisfactory and is unlikely to improve (e.g. K(s) has enormous gains), then abandon the current design and restart with step 1.

(3) With the precompensator K(s) determined in the previous step, use Algorithm 7.2.2 to solve for (L,F). Study the compensated transmittance and either (a) stop, (b) go on to step 2, or (c) restart with step 1 according to (i), (ii) and (iii) of step 2 above.

End of Design Procedure 7.2.3

We do not claim that this procedure always works. A characteristic feature of the design methodology outlined above however is that it offers many flexible options. If one is willing to iterate a number of times, adjusting various specifications during the process, then Algorithms 7.2.1, 7.2.2 and Design Procedure 7.2.3 may be regarded as a powerful computational means for achieving the designer's goals.

§7.3 Design Examples for Systems with More Outputs than Inputs

Three examples will now be given for Design Procedure 7.2.3. Each example illustrates one of the possible starting options given in that procedure.

Example 7.3.1

Consider the non-square model NSRE (see Appendix D) of a chemical reactor which is open-loop unstable. Naturally, in this case our first objective is to stabilize the plant. We do this by an inner feedback loop; that is, we start off Design Procedure 7.2.3 by option 1(i).

For the inner feedback loop, we specify a reasonably modest target of using the pair (L,F) to balance up high frequency gains as well as stabilizing the plant. The specifications are given in Fig.7.8(a-e). Algorithm 7.2.2 then produced

$$L = \begin{bmatrix} 0.0128 & 0.444 \\ -0.743 & -0.471 \end{bmatrix}$$

$$F = \begin{bmatrix} 1.24 & 1.91 & -2.29 \\ -0.207 & -0.540 & -0.0670 \end{bmatrix}$$

The corresponding graphs, for the compensated $H(s)=G_1(s)L\left(I+FG(s)L\right)^{-1}$, are given in Fig.7.9(a-h). The inner-loop compensated H(s) is now stable having poles at $\{-0.621, -0.843, -5.32, -9.86\}$.

Next, a PI-precompensator K(s) is constructed by fitting H(s)K(s) to $\frac{10}{s}I_2$, with respect to a weighting $|(100s+1)/(s+10)|^2 \mathbb{1}_2$. The result obtained was

$$K(s) = \frac{1}{s}\begin{bmatrix} 3.20\ s + 1.89 & -0.374\,s + 1.02 \\ -0.0230\,s + 1.65 & 3.51\,s + 3.34 \end{bmatrix}$$

K(s) has zeros at $\{-0.323, -1.28\}$. The precompensated system H(s)K(s) has the characteristics given in Fig.7.10, which are regarded as satisfactory, and hence we conclude this design by noting that we have gone through the sequence 1(i), 3(b), 2(i) of Design Procedure 7.2.3.

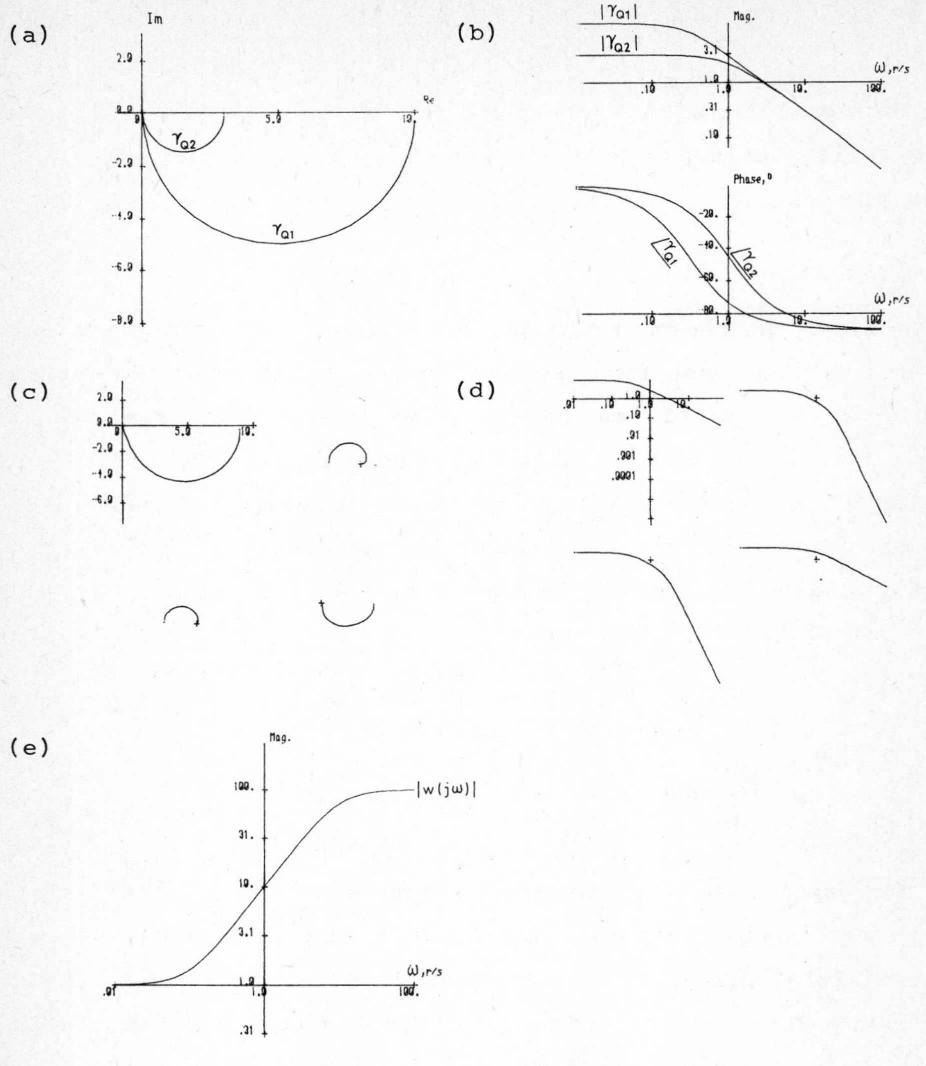

<u>Fig.7.8</u> Specifications for calculating (L,F) for NSRE.

(a),(b) The desired characteristic gains are:

$$\gamma_{Q1}(s) = \frac{3}{s+0.3}\,, \quad \gamma_{Q2}(s) = \frac{3}{s+1}$$

(c),(d) Nyquist and Bode magnitude arrays of the desired response.

(e) The weighting function $|\frac{100s+10}{s+10}|$ for inner loop fitting.

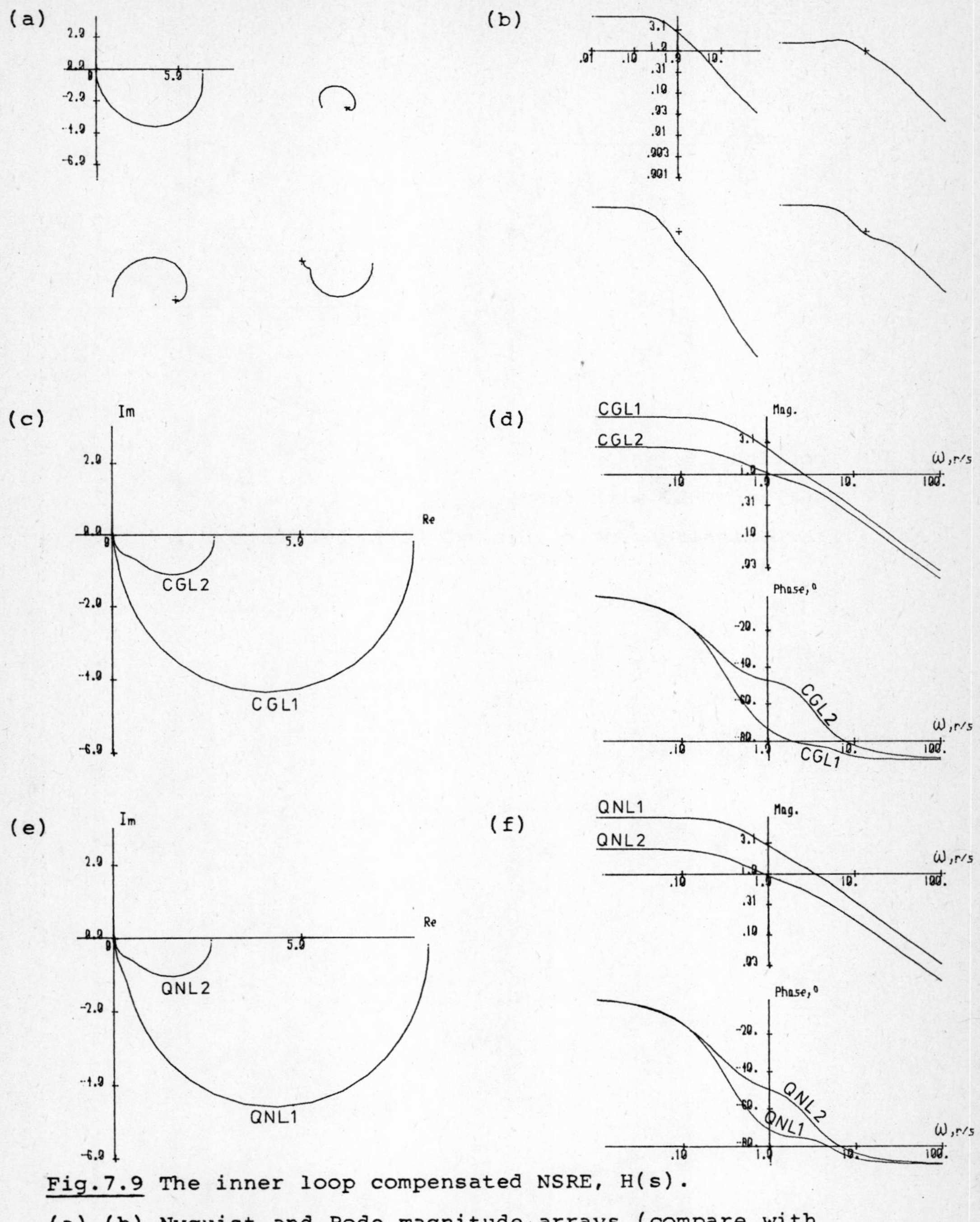

<u>Fig.7.9</u> The inner loop compensated NSRE, H(s).

(a),(b) Nyquist and Bode magnitude arrays (compare with Fig.7.8(c,d)).

(c),(d) Characteristic gain loci (compare with Fig.7.8(a,b)).

(e),(f) QN loci.

(g)

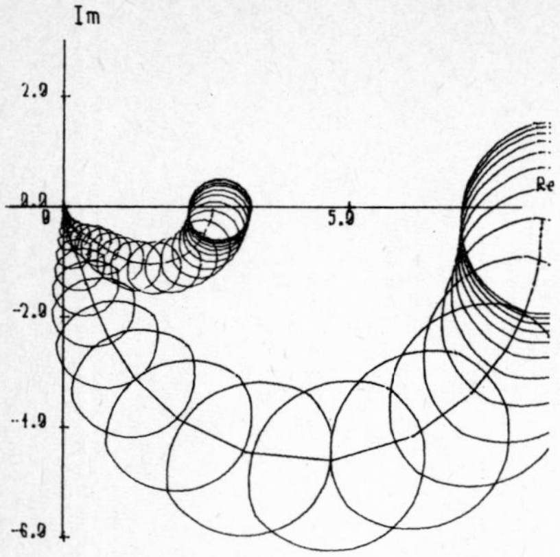

(h)

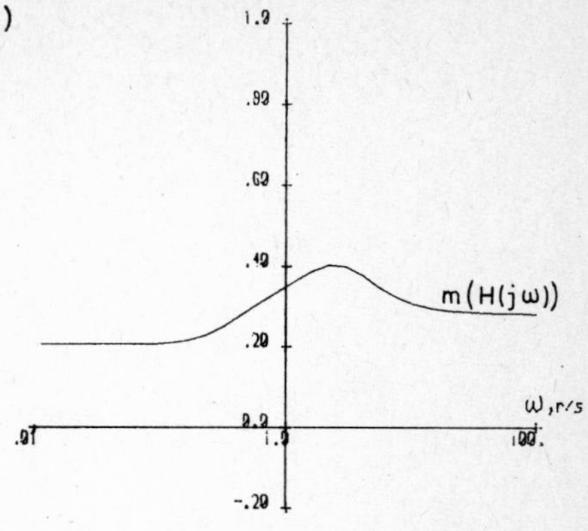

Fig.7.9 (continued)

(g) QNL1 and QNL2 with bands.

(h) Frame misalignment of $H(s)=G_1(s)L[I+FG(s)L]^{-1}$.

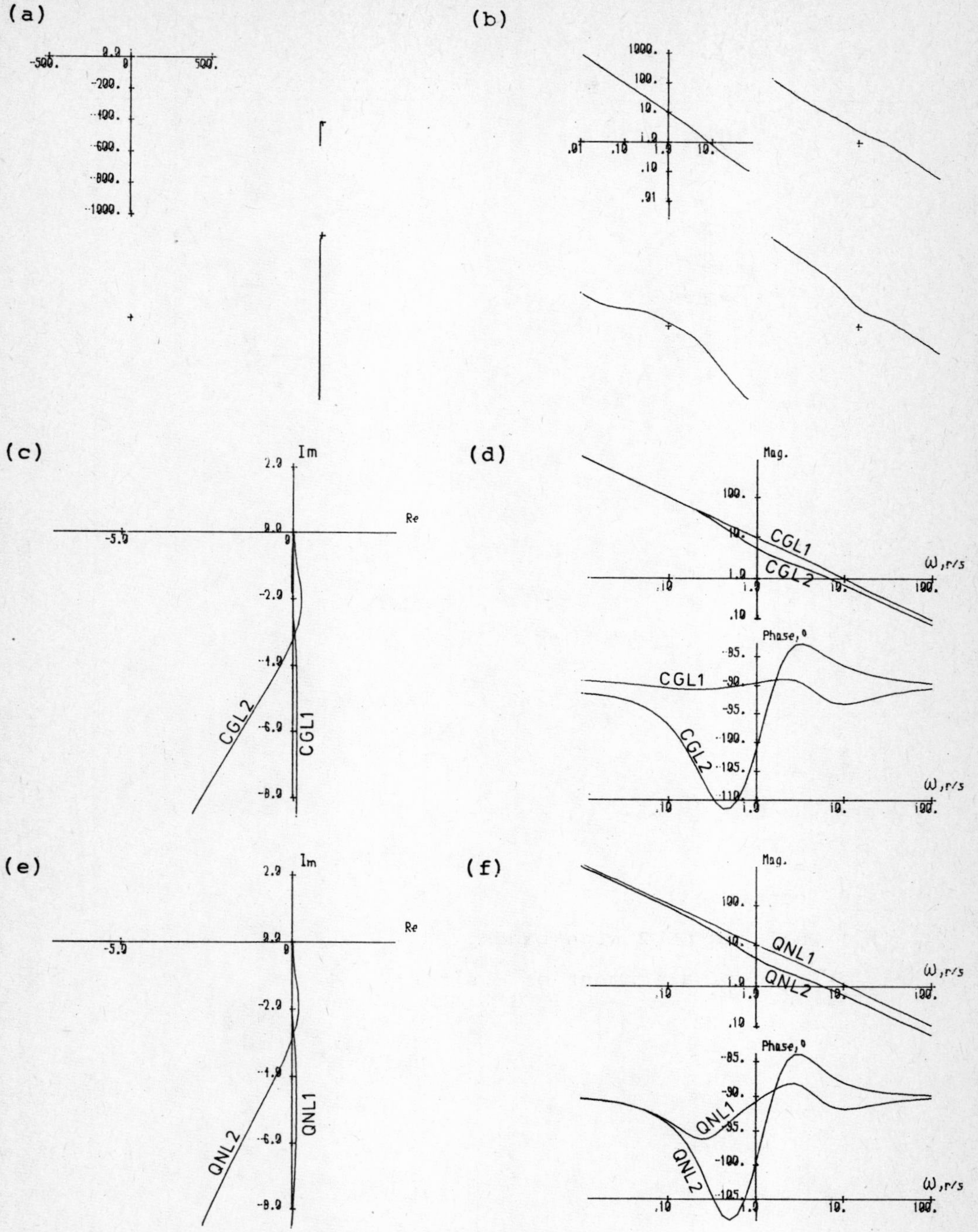

Fig.7.10 The final compensated NSRE, H(s)K(s).

(a),(b) Nyquist and Bode magnitude arrays (compare with $\frac{10}{s}I_2$).

(c),(d) Characteristic gain loci.

(e),(f) QN loci.

(g)

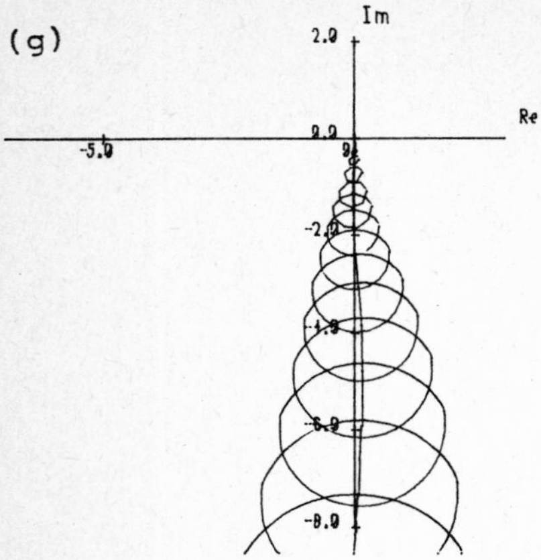

(h)

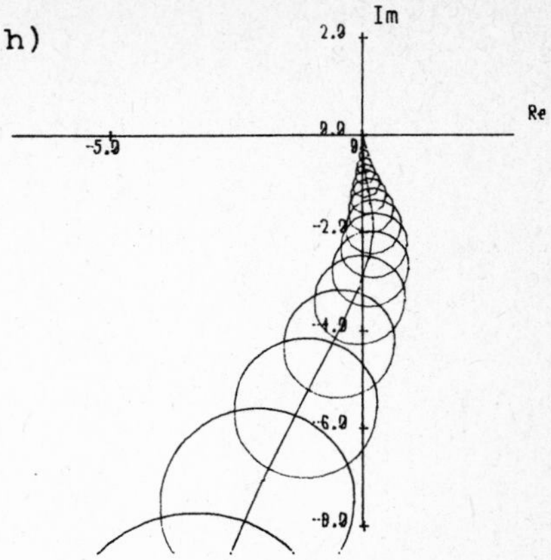

(i)

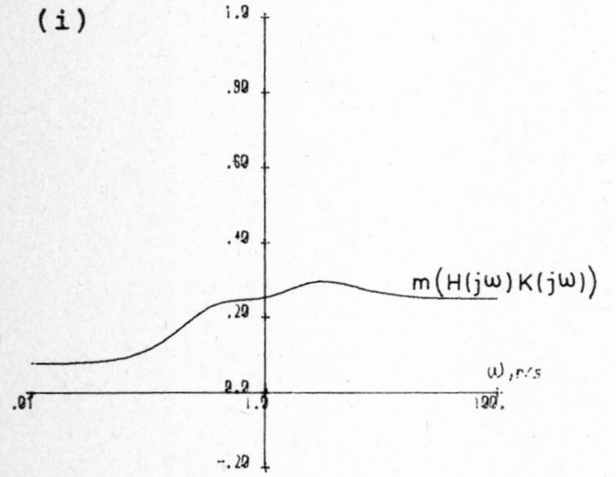

<u>Fig.7.10</u> (continued)

(g),(h) QNL1 and QNL2 with bands.

(i) Frame misalignment of H(s)K(s).

Example 7.3.2

The system, AIRC, to be considered next is a state-space model of the linearized vertical-plane dynamics of an aircraft, details of which are given in Appendix F.

AIRC has 3 inputs and 5 states, all of which are measured. We investigate the design of an automatic flight control system, having the configuration of Fig.7.6, to control the first 3 states.

First, we make a few basic specifications which will be used throughout the design procedure:

(i) All calculations are based on a logarithmic equally spaced frequency list $\{\omega_1,\ldots,\omega_n\}$ of 50 points.

(ii) The desired characteristic gain loci, in $\Gamma_Q(s) = \text{diag}\{\gamma_{Qi}(s)\}$, are chosen to be

$$\gamma_{Q1}(s) = \frac{2}{s} \quad , \quad \gamma_{Q2}(s) = \gamma_{Q3}(s) = \frac{6}{s(s+3)}$$

Nyquist and Bode plots of $\gamma_{Qi}(j\omega)$ (i=1,2,3) are given in Fig.7.11(a,b). The corresponding desired responses, defined as $Y_\rho \Gamma_Q(j\omega_\rho) Y_\rho^*$ (see (7.2.9) and (7.2.10)), are given in Fig.7.11(c,d).

(iii) The weighting matrix used for the fitting Algorithms 7.2.1 and 7.2.2 is $W(s) = |w(s)|^2 \mathbb{1}_3$ where the weighting function $|w(j\omega)|$ is shown in Fig.7.11(e).

We can now proceed with Design Procedure 7.2.3:

(1) Start by setting $L = I$ and $F = O$. This means that initially, the inner loop is open and the measurements available in the extra outputs are ignored.

(2) Using Algorithm 7.2.1, a precompensator K(s) is designed for $G_1(s)$, the upper square block of G(s). (Note that H(s) of (7.2.8) reduces to $G_1(s)$ for $(L,F) = (I,O)$.) Since all the states of the system are already accessible, we do not envisage the use of a complicated controller. Hence we restrict K(s) to a PI-precompensator of the

form

$$K(s) = \frac{1}{s}N(s)$$

with

$$\deg(n_{ij}(s)) = 1 \quad i,j = 1,2,3$$

Solving (7.2.9) gave

$$N(s) = \begin{bmatrix} -1.46\,s - 0.0345 & -0.0088\,s + 0.411 & -0.183\,s + 0.861 \\ -0.0454\,s - 0.00582 & 1.88\,s + 0.108 & -0.106\,s + 0.160 \\ -2.04\,s - 0.0361 & -0.0228\,s + 1.13 & -1.08\,s + 1.16 \end{bmatrix}$$

Graphs for the resulting $G_1(s)K(s)$ are given in Fig.7.12. The characteristic gain loci (c,d) show that the closed-loop system will be only just stable. The principal gain plot (e) shows that the principal gains are unbalanced and fall short of the desired bandwidth.

(3) Fixing the precompensator K(s) just calculated, we now go on to tune (L,F) of the inner loop. Restricting L to be a diagonal matrix and solving (7.2.10) gave

$$L = \text{diag}\{\, 5.16\,, 1.00\,, 8.91\,\}$$

$$F = \begin{bmatrix} -0.0158 & 0.171 & -0.379 & -0.0137 & 0.633 \\ -0.00270 & 0.00003 & -0.117 & -0.0350 & 0.0891 \\ -0.0173 & 0.526 & -0.553 & -0.380 & 0.942 \end{bmatrix}$$

from a starting point of $(L_0, F_0) = (I, O)$. The "open-loop" compensated system is now $G_1(s)L(I + FG(s)L)^{-1}K(s)$, with accompanying graphs given in Fig.7.13, which look satisfactory.

It was presumed that the inner loop would be stable and it is necessary to check that this is so, for the otherwise nicely fitted characteristic gain loci may give a wrong number of encirclements around the critical point. Standard calculations show that the inner loop $G_1(s)L(I+FG(s)L)^{-1}$ has poles at

$$\{ -0.00024, -0.0460 \pm j0.203, -5.68 \pm j0.617 \}$$

all in the LHP, as required for stability. It is also instructive to see where the zeros of K(s) are located. They turn out to be at

$$\{ -0.00016, -0.0458 \pm j0.203 \}$$

Comparing these with the pole positions of the plant AIRC suggests that the minimization algorithm first placed the zeros of K(s) fairly close to the subset of poles $\{0, -0.018 \pm j0.182\}$ of AIRC (see Appendix F). It is, of course, to be expected that K(s) will at least try to cancel the pair of resonant modes $-0.018 \pm j0.182$ of AIRC. Next, in step (3), the inner loop adjusted itself in such a way that the inner loop had poles practically cancelling the zeros of K(s). The other two poles were shifted to $-5.68 \pm j0.617$ to increase the bandwidth.

We conclude this example by noting that we have gone through the sequence 1(ii), 2(ii) and 3(a) of Design Procedure 7.2.3.

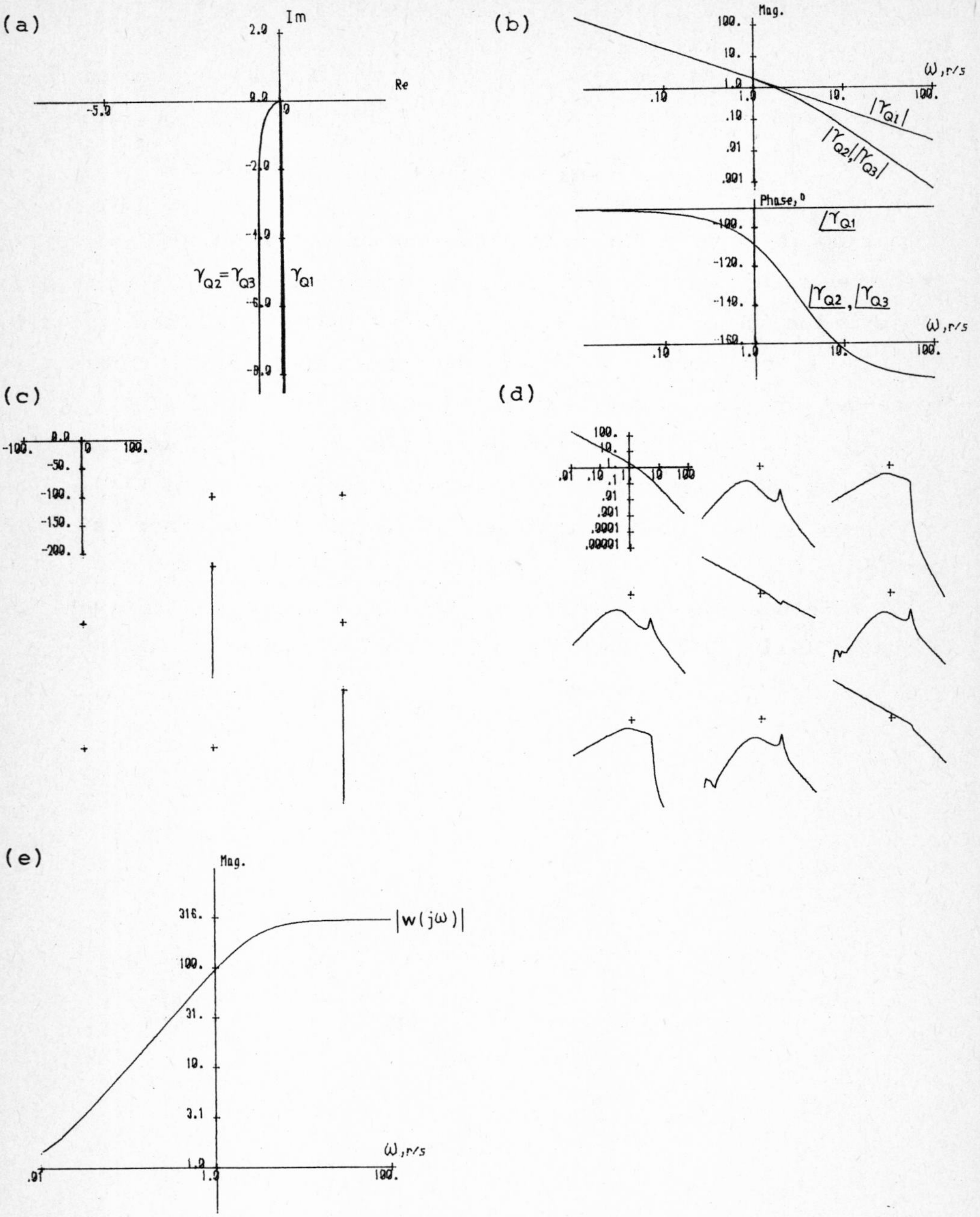

Fig.7.11 Specifications for the fitting algorithms (Examples 7.3.2 & 7.3.3).

(a),(b) The desired characteristic gain loci $\gamma_{Qi}(j\omega)$, i=1,2,3.

(c),(d) Nyquist and Bode magnitude arrays of the desired response.

(e) The weighting function is $\left|\frac{300s+3}{s+3}\right|$.

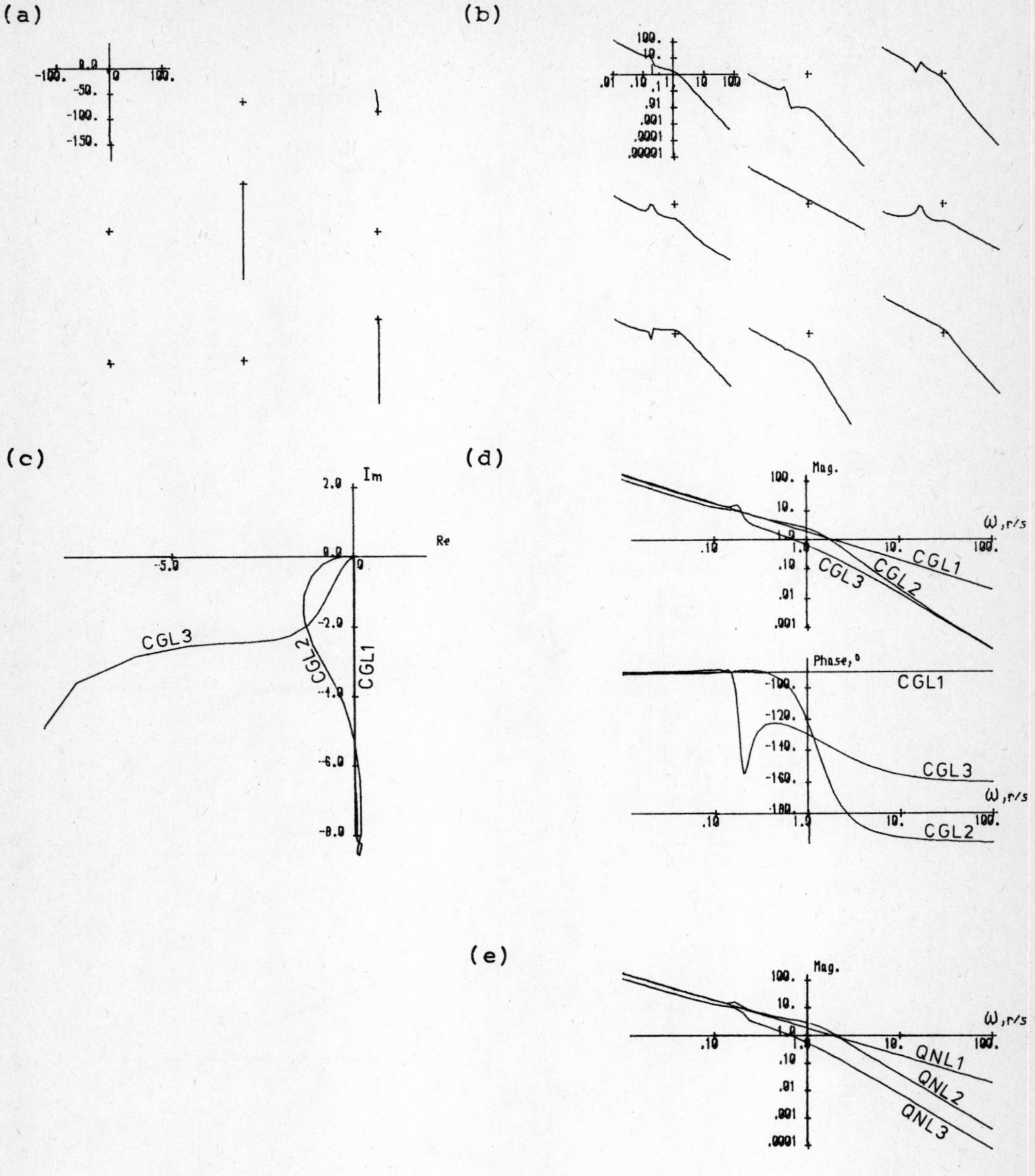

<u>Fig.7.12</u> Step (2) of Example 7.3.2 yields the system $G_1(s)K(s)$.

(a),(b) Nyquist and Bode magnitude arrays.

(c),(d) Characteristic gain loci.

(e) Principal gains (i.e., QN loci in Bode magnitude form).

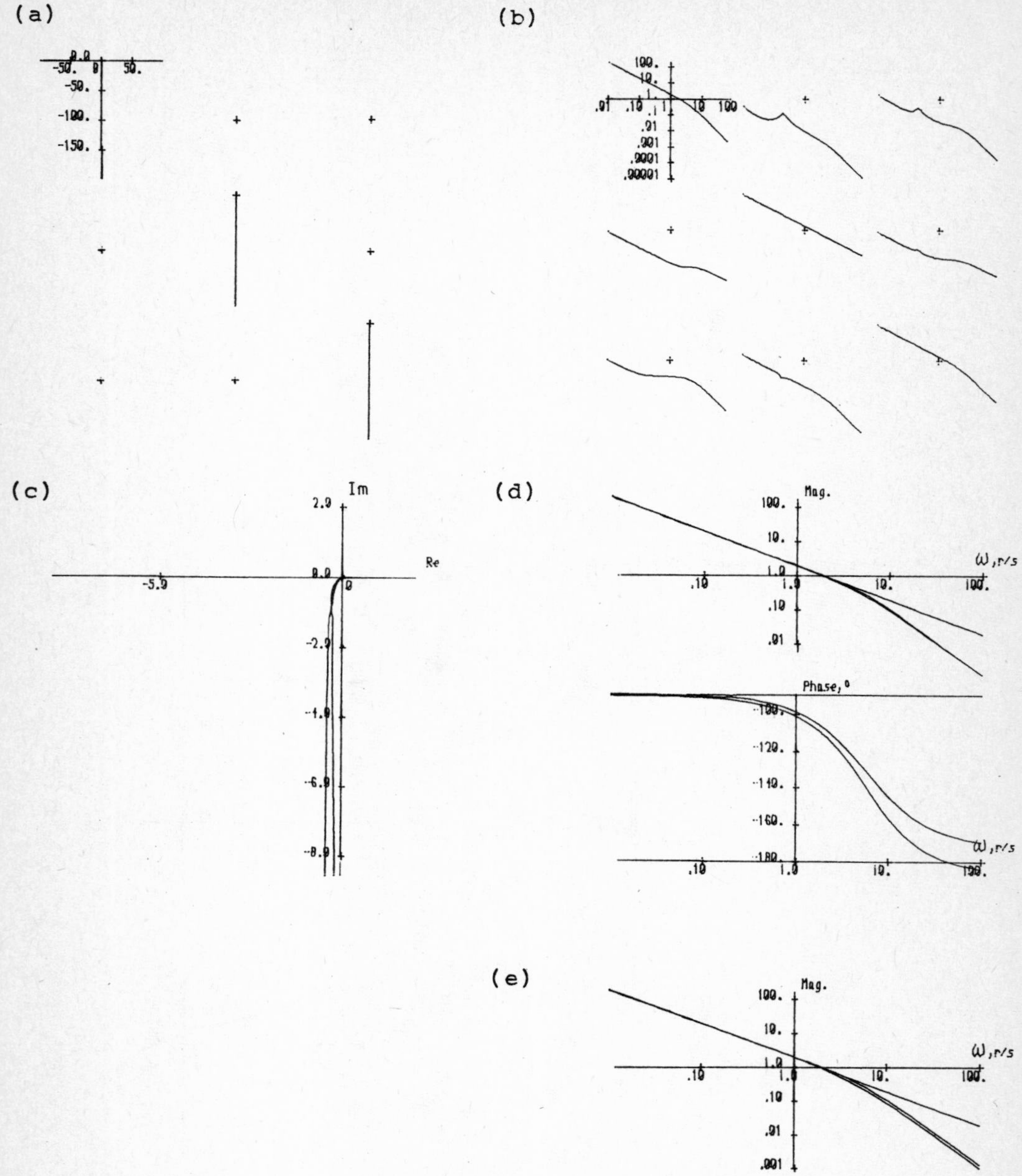

Fig.7.13 Step (3) of Example 7.3.2 yields the system $G_1(s)L[I+FG(s)L]^{-1}K(s)$.

(a),(b) The Nyquist and Bode magnitude arrays show that the system is nearly diagonalized.

(c),(d) The characteristic gain loci show that the system will be stable with the negative-unity loop closed.

(e) The principal gains are well balanced.

Example 7.3.3

In this example, we shall design another automatic flight controller for the same system AIRC which was considered in Example 7.3.2. For the sake of comparison, we retain the same basic specifications as before, including the frequency list, $\gamma_{Qi}(s)$ (i=1,2,3) and the weighting matrix (see Fig.7.11). We now start Design Procedure 7.2.3 by taking the option 1(iii):

(1) Since all 5 states of AIRC are available, we can readily construct a Linear-Quadratic Regulator (LQR) for it by solving for the positive-definite P in the standard algebraic Riccati equation (e.g. see [KAI, pp.230])

$$PA + A^T P + PBR^{-1} B^T P - Q = 0 \tag{7.3.1}$$

where $R = R^T$ and $Q = Q^T$ are appropriate weighting matrices for the control inputs and the states respectively. Putting

$$L = I \tag{7.3.2}$$

$$F = -R^{-1} B^T P \tag{7.3.3}$$

will then make the inner loop of Fig.7.7 an LQR. Since (L,F) will be re-adjusted later, we shall not be too fussy in the choices of Q and R. We simply note that, to keep the initial feedback gains small, we can set R to be relatively larger than Q. So we put

$$Q = \text{diag}\{ 1,1,1,0,0 \}$$

$$R = 10\, I_3$$

where Q weights the controlled variables only. (7.3.1) and (7.3.3) then give

$$F = \begin{bmatrix} 0.0723 & 0.121 & 0.695 & 0.268 & -0.343 \\ 0.158 & 0.434 & 0.490 & 0.138 & -0.308 \\ -0.264 & -0.208 & -0.952 & -0.289 & 0.543 \end{bmatrix} \tag{7.3.4}$$

(2) Now fix the pair (L,F) given by (7.3.2) and (7.3.4). Restricting K(s) to be a PI-precompensator and running through Algorithm 7.2.1, we get

$$K(s) = \frac{1}{s}N(s) \tag{7.3.5}$$

where

$$N(s) = \begin{bmatrix} -1.45\ s + 0.268 & -0.0560\ s + 0.655 & -0.0191\ s + 1.56 \\ -0.0724\ s + 0.380 & 1.47\ s + 1.01 & 0.0605\ s + 0.445 \\ -2.94\ s - 0.406 & -0.133\ s + 0.767 & -2.08\ s + 0.723 \end{bmatrix} \tag{7.3.6}$$

At this point, the transmittance $T_{ec}(s)$ (see (7.2.7)) appears as shown in Fig.7.14. Clearly it is necessary to adjust the inner loop to get closer to the specified target shown in Fig.7.11.

(3) Fixing the K(s) just calculated and restricting L to be diagonal, Algorithm 7.2.2 gives

$$L = \text{diag}\{\ 4.07, 1.03, 4.40\ \} \tag{7.3.7}$$

$$F = \begin{bmatrix} 0.136 & 0.279 & 0.0899 & 0.0824 & 0.511 \\ 0.191 & 0.438 & 0.105 & 0.0351 & 0.0272 \\ -0.207 & 0.249 & -1.31 & -0.883 & 1.34 \end{bmatrix} \tag{7.3.8}$$

With the K(s), (L,F) now defined by (7.3.5-8), the transmittance $T_{ec}(s)$ has the characteristics shown in Fig.7.15, which look satisfactorily close to the desired response.

Further analysis shows that the inner loop is stable with poles at $\{-0.493, -0.534 \pm j0.223, -4.01, -7.26\}$ and that K(s) has zeros at $\{-0.572, -0.534 \pm j0.272\}$. Again it is apparent that the poles of the inner loop are approximately cancelling the zeros of the precompensator.

We conclude this example by noting that we have gone through the sequence 1(iii), 2(ii) and 3(a) of Design Procedure 7.2.3.

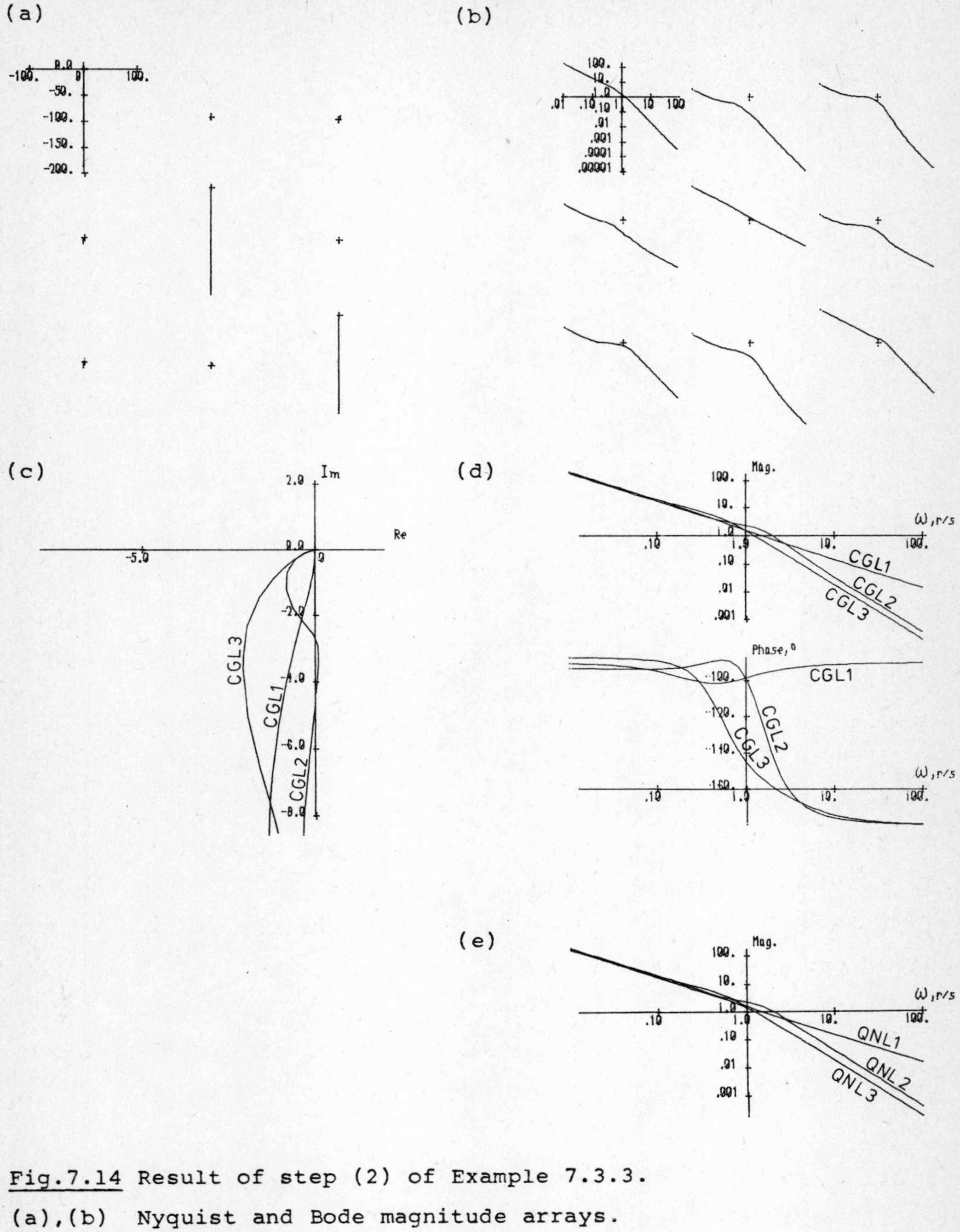

Fig.7.14 Result of step (2) of Example 7.3.3.

(a),(b) Nyquist and Bode magnitude arrays.

(c),(d) Characteristic gain loci.

(e) Principal gains.

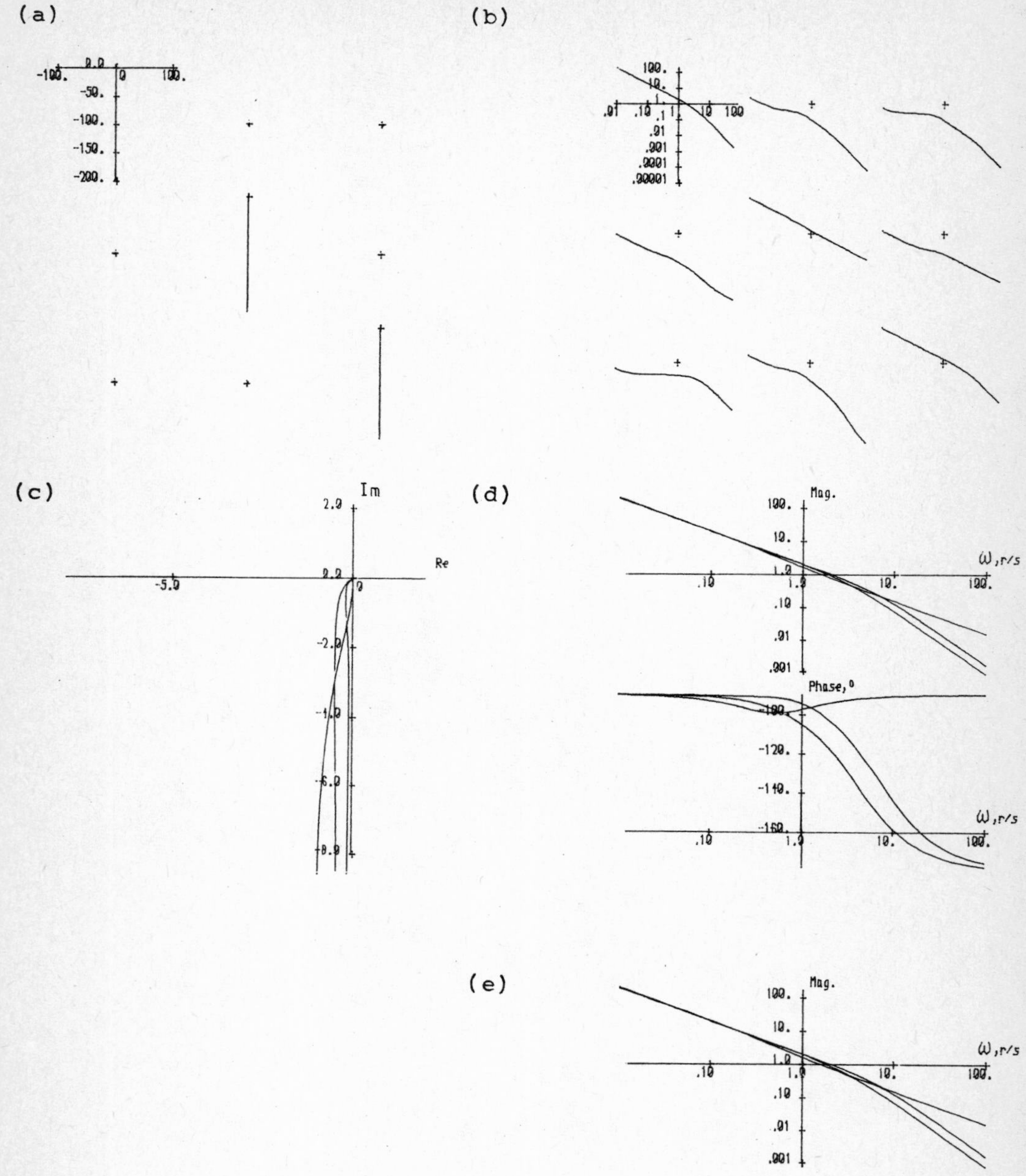

<u>Fig.7.15</u> Result of step (3) of Example 7.3.3.

(a),(b) Nyquist and Bode magnitude arrays.

(c),(d) Characteristic gain loci.

(e) Principal gains.

As far as the transmittance $T_{ec}(s)$ is concerned, the two designs given in Examples 7.3.2 and 7.3.3 gave very much the same results (compare Fig.7.13 and Fig.7.15). In order to further evaluate the two controllers, we can investigate the robustness property, at the plant input, of the closed-loop configuration Fig.7.6. Partitioning $F = [F_1 \mid F_2]$ compatibly with $G(s) = \begin{bmatrix} G_1(s) \\ \hline G_2(s) \end{bmatrix}$ and breaking the loops at the plant input gives the return-ratio matrix

$$L[K(s)+F_1 \mid F_2] \begin{bmatrix} G_1(s) \\ \hline G_2(s) \end{bmatrix}$$

The corresponding stability margins (see §4.4) for the results of Examples 7.3.2 and 7.3.3 are shown in Fig.7.16. They show that the design of Example 7.3.3 is to be preferred.

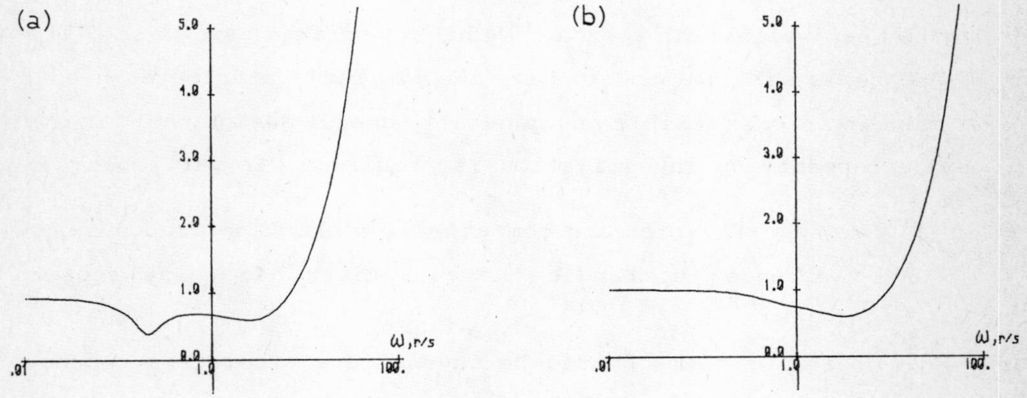

Fig.7.16 Stability margins at the plant input

i.e. $\sigma_{min}\left\{ I + \left(L[K(j\omega)+F_1 \mid F_2] \begin{bmatrix} G_1(j\omega) \\ \hline G_2(j\omega) \end{bmatrix} \right)^{-1} \right\}$

for (a) Example 7.3.2 and (b) Example 7.3.3

§7.4 General Conclusion

The computer-aided design of feedback controllers is likely to become a widespread practice as the power and availability of interactive computing systems increases. Whatever particular techniques and approaches are to be used it seems clear that any useful CAD package will have the following two basic attributes:

(i) The analysis aspect will be conceived in terms of a conceptual framework which is congenial to the practising engineer.

(ii) The synthesis aspect will remove as much as possible of the burden of detailed parameter manipulation from the designer, leaving him free to concentrate on trading off his desired specifications against the price which has to be paid to achieve them.

Classical feedback theory, by which we mean the frequency-response approach of Nyquist and Bode, can be extended to provide a conceptual framework within which to handle multivariable feedback systems. The Quasi-classical approach we have described here is in this spirit and seems well adapted to have the two basic attributes needed. To make it into a flexible and generally useful design tool some further development and investigation is required. In particular:

(i) the specification and compatibility checking precedures should be extended to handle the situation when noisy sensors are present;

(ii) digital systems should be investigated from this approach;

(iii) the flexibility of being able to optimize over a range of specified controller structures should be investigated further; and

(iv) the trade-offs between normalization, robustness, interaction and controller complexity deserve further study.

In general terms however we feel that the results presented here show that the Quasi-classical approach enables the multivariable feedback control design problem to be handled in a way which is well adapted to computer-aided-design by practising engineers. In particular, the combination of singular values with appropriate phase information enables the classical Nyquist-Bode stability and performance indicators to be used to handle all three key aspects of the feedback problem: stability, performance and robustness.

APPENDIX A ANALYTIC PROPERTIES OF THE SINGULAR VALUES OF A RATIONAL MATRIX

§A.1 Analytic Properties of the Characteristic Values

A rational matrix $G(s) \varepsilon \mathbb{R}(s)^{m\times m}$ has an associated set of frequency-dependent eigenvalues $\{g_i(s) \mid i=1,\ldots,m\}$, usually called the characteristic gains. If G(s) has a right coprime MFD

$$G(s) = N(s)D(s)^{-1} \tag{A.1.1}$$

then these are determined from the equation

$$\Delta(s,g) := \det[gD(s) - N(s)] = 0$$

where $\Delta(s,g)$ is unique up to a possible constant for all possible right coprime factorizations of G(s). $\Delta(s,g)$ can be factored into:

$$\Delta(s,g) = e(s)\Delta_1(s,g)\Delta_2(s,g)\cdots\Delta_q(s,g)c(g)$$

where $\{\Delta_i(s,g) \mid i=1,2,\ldots,q\}$ are the irreducible factors of $\Delta(s,g)$ dependent on both s and g, and the other factors represent factors independent of s (c(g)) or g (e(s)). If we assume, for simplicity of exposition,

$$e(s) = 1 \quad \text{and} \quad c(g) = 1$$

and take $\Delta(s,g)$ to be irreducible, and of the form

$$\Delta(s,g) = a_m(s)g^m + a_{m-1}(s)g^{m-1} + \cdots + a_o(s) \tag{A.1.2}$$

then this (implicitly) defines g as an <u>algebraic function</u> of s. For all s such that $a_m(s) \neq 0$, (A.1.2) defines an m-valued relation:

$$g(s) : \mathbb{C} - \{s \varepsilon \mathbb{C} \mid a_m(s) = 0\} \rightarrow \mathbb{C}$$

If $\Omega \subset \mathbb{C}$ is a domain which excludes all singular points (poles and branch points) then for $s \varepsilon \Omega$ the characteristic gain functions $\{g_i(s) \mid i=1,2,\cdots,m\}$ are (complex) analytic functions of s. In global terms these can be organized into a "single" analytic function g(s) — the characteristic gain function — whose domain is a Riemann surface. For further information on the analytic properties of the characteristic gain functions and the background algebraic function theory, the reader may consult [POS1],[SMI1],[BLI].

§A.2 Analytic Properties of the Singular Values

Let $G(s) \in \mathbb{R}(s)^{m \times \ell}$. For notational simplicity, we assume $m \geq \ell$ throughout this section. Let G(s) have a right coprime MFD given by

$$G(s) = N(s)D(s)^{-1} \tag{A.2.1}$$

Then the squares of the singular values, that is the eigenvalues of $G(s)^*G(s) = G(\bar{s})^T G(s)$, are determined by solving the equation

$$\det\left[\sigma^2 D(\bar{s})^T D(s) - N(\bar{s})^T N(s)\right] = 0$$

and one could attempt to investigate the analytic properties of the frequency-dependent singular values by considering a "squared-singular-value function" $\sigma^2(s,\bar{s})$ defined implicitly via the algebraic equation

$$\chi(s,\bar{s},\sigma^2) := \det\left[\sigma^2 D(\bar{s})^T D(s) - N(\bar{s})^T N(s)\right] = 0$$

This is not a very satisfactory approach, and, in order to explore the use of existing results for functions of several complex variables, it turns out to be better to consider the equation in three indeterminates s_1, s_2 and σ^2 given by

$$\chi(s_1,s_2,\sigma^2) := \det\left[\sigma^2 D(s_2)^T D(s_1) - N(s_2)^T N(s_1)\right] = 0 \tag{A.2.2}$$

This frees us from the awkwardness of handling the conjugate term, and the investigation proceeds first over the two complex variables s_1 and s_2, then specializes down to the required result by putting $s_1 = s$ and $s_2 = \bar{s}$. Before proceeding further it will be helpful to give some necessary background definitions and results.

§A.2.1 Analytic Sets, Pseudopolynomials and Real-Analytic Funtions

An analytic set is a subset of $\mathbb{C}^n$ defined locally by the zeros of a set of holomorphic functions. Let Ω be an open set in $\mathbb{C}^n$. Then $A \subset \Omega$ is called an analytic set (e.g. see [NAR, pp.50]) if $\forall a \in A$, $\exists$ a neighbourhood N of a and finitely many holomorphic functions $f_1, f_2, \ldots, f_p$ in N s.t.

$$N \cap A = \{z \in N \mid f_1(z) = f_2(z) = \cdots = f_p(z) = 0\}$$

Let $z := (z_1, \ldots, z_n) \in \mathbb{C}^n$. Then a function

$$f(z,\nu) := \nu^n + a_1(z)\nu^{n-1} + \cdots + a_n(z) \qquad (A.2.3)$$

is called a pseudopolynomial when the coefficients $\{a_i(z) \mid i = 1, \ldots, n\}$ are holomorphic functions of z on some open set $\Omega \subset \mathbb{C}^n$. If Ω is connected, then the set of all holomorphic functions on Ω is an integral domain. Moreover, one can consider factorizing the pseudopolynomial, regarded as a polynomial in ν, into a set of factors unique over this integral domain (e.g. see [GRA, Chapter 3 §6]). Furthermore $f(z,\nu)$ will have an associated pseudopolynomial discriminant function $D_\nu(f)$. At any z' for which $D_\nu(f)$ vanishes, $f(z',\nu)$ will have repeated roots in ν. The totality of these, defined as the zero set of the discriminant function, is an analytic set.

We can now state the key result we require (see [GRA, Chapter III Theorem 6.12]).

Theorem A.2.1

Let $\Omega \subset \mathbb{C}^n$ be a domain and let $f(z,\nu)$ be a pseudopolynomial in ν whose discriminant $D_\nu(f)$ does not vanish identically in z. Then the sets of zeros

$$Z(f) := \{(z,\nu) \in \Omega \times \mathbb{C} \mid f(z,\nu) = 0\}$$

$$Z(D_\nu(f)) := \{z \in \Omega \mid D_\nu(f) = 0\}$$

are both analytic sets.

For any $z^\circ \in \Omega - Z(D_\nu(f))$ there exists an open neighbourhood of z°,

N, and a set of holomorphic functions $\{\nu_1, \nu_2, \ldots, \nu_n\}$ on N, with $\nu_i(z) \neq \nu_j(z)$ for $i \neq j$ and $\forall z \in N$ s.t.

$$f(z,\nu) = (\nu - \nu_1(z))(\nu - \nu_2(z)) \cdots (\nu - \nu_n(z)) \qquad \forall z \in N$$

□

Finally we need to define the term real-analytic.

Let $\Omega \subset \mathbb{R}^n$ be an open set, and let $f(x)$ be a real-valued function of $x = (x_1, \ldots, x_n) \in \Omega$. Then $f(x)$ is said to be real-analytic at $x^\circ = (x_1^\circ, \ldots, x_n^\circ) \in \Omega$ if there exists a neighbourhood N of x° and a set of real coefficients $c_\alpha \in \mathbb{R}$ s.t.

$$f(x) = \sum_\alpha c_\alpha (x - x^\circ)^\alpha$$

where $\alpha = (\alpha_1, \ldots, \alpha_n)$ is an n-tuple of non-negative integers and we have used the standard notation that

$$(x - x^\circ)^\alpha := (x_1 - x_1^\circ)^{\alpha_1} \cdots (x_n - x_n^\circ)^{\alpha_n}$$

If f can be expanded as a convergent Taylor series for every point in Ω, then $f(x)$ is said to be real-analytic in Ω. Real analytic functions possess derivatives of all orders and the principle of analytic continuation may be applied to them (e.g. see [NAR, pp.4] or [JOH, pp.65]).

§A.2.2 Real-Analyticity of Singular-Value Functions

We now proceed to demonstrate that locally the frequency-dependent singular values $\{\sigma_i(s,\bar{s}) \mid i = 1,2,\ldots,\ell\}$ of a rational matrix $G(s) \in \mathbb{R}(s)^{m \times \ell}$ are real-analytic functions of x and y, where $s = x + jy$. Let G(s) have a right coprime MFD given by (A.2.1) and consider the polynomial equation in the indeterminates s_1, s_2 and σ^2 given by (see (A.2.2)):

$$\det[\sigma^2 D(s_2)^T D(s_1) - N(s_2)^T N(s_1)] = 0 \qquad (A.2.4)$$

Expanding the determinant we get

$$p(s_2) p(s_1) \sigma^{2\ell} + \cdots + (-1)^\ell z(s_2) z(s_1) = 0 \qquad (A.2.5)$$

where $p(\cdot) := \det[D(\cdot)]$ and $z(\cdot) := \det[N(\cdot)]$. On dividing through (A.2.5) by the leading-term's coefficient, we get

$$f(s_1, s_2, \sigma^2) := \sigma^{2\ell} + \cdots + (-1)^{\ell} \frac{z(s_2)z(s_1)}{p(s_2)p(s_1)} = 0 \qquad (A.2.6)$$

which is a pseudopolynomial in σ^2.

Now let Ω be an open set given by

$$\Omega := (\mathbb{C} - \{\text{zeros of } p(\cdot)\}) \times (\mathbb{C} - \{\text{zeros of } p(\cdot)\})$$

so that the coefficients of the pseudopolynomial $f(s_1, s_2, \sigma^2)$ are holomorphic in s_1 and s_2 over Ω. If we assume that the discriminant $D_\nu(f)$ (here ν stands for σ^2) of the pseudopolynomial $f(s_1, s_2, \sigma^2)$ does not vanish identically in Ω then Theorem A.2.1 above tells us that

$$Z(f) = \{(s_1, s_2, \sigma^2) \in \Omega \times \mathbb{C} \mid f(s_1, s_2, \sigma^2) = 0\}$$

and $$Z(D_\nu(f)) = \{(s_1, s_2) \in \Omega \mid D_\nu(f) = 0\}$$

are analytic sets. Furthermore, for any (s_1°, s_2°) s.t.

$$(s_1^\circ, s_2^\circ) \in \Omega - Z(D_\nu(f))$$

there will exist a neighbourhood N of (s_1°, s_2°) on which $f(s_1, s_2, \sigma^2)$ factors <u>locally</u> into distinct linear factors:

$$f(s_1, s_2, \sigma^2) = (\sigma^2 - \sigma_1^2(s_1, s_2)) \cdots (\sigma^2 - \sigma_\ell^2(s_1, s_2)) \qquad (A.2.7)$$

where each term $\sigma_i^2(s_1, s_2)$ is holomorphic for all $(s_1, s_2) \in N$. Consequently, in the neighbourhood N, the holomorphic functions $\sigma_i^2(s_1, s_2)$ — which are the "solutions" of the equation (A.2.6) — can be expanded as a Taylor series. For notational simplicity, let us take the point (s_1°, s_2°) to be simply the point (0,0). The Taylor series then take the form (e.g. see [GRA, Chapter I Theorem 3.8]):

$$\sigma_i^2(s_1, s_2) = \sum_{\alpha_1, \alpha_2} d_{i\alpha_1\alpha_2} s_1^{\alpha_1} s_2^{\alpha_2} \qquad i = 1, 2, \ldots, \ell \qquad (A.2.8)$$

where the coefficients $d_{i\alpha_1\alpha_2}$ are given by

$$d_{i\alpha_1\alpha_2} = \frac{1}{\alpha_1!\,\alpha_2!} \frac{\partial^{\alpha_1+\alpha_2} \sigma_i^2(s_1,s_2)}{\partial s_1^{\alpha_1} \partial s_2^{\alpha_2}} \tag{A.2.9}$$

We may now recover the squared-singular-value functions $\{\sigma_i^2(s,\bar{s}) \mid i=1,2,\ldots,\ell\}$ of G(s) by making the substitutions $s_1 = s$ and $s_2 = \bar{s}$. If $s = x + jy$ then we can regard $\sigma_i^2(s,\bar{s})$ as a function of the real variables $(x,y) \in \mathbb{R}^2$. With a mild abuse of notation, we shall write $\sigma_i^2(x+jy,x-jy)$ as $\sigma_i^2(x,y)$. Accordingly, let us rewrite (A.2.8) as

$$\begin{aligned}\sigma_i^2(x,y) &= \sum_{\alpha_1,\alpha_2} d_{i\alpha_1\alpha_2} (x+jy)^{\alpha_1} (x-jy)^{\alpha_2} \\ &= \sum_{\beta_1,\beta_2} (e_{i\beta_1\beta_2} + jh_{i\beta_1\beta_2})\, x^{\beta_1} y^{\beta_2}\end{aligned}$$

where $e_{i\beta_1\beta_2}, h_{i\beta_1\beta_2} \in \mathbb{R}$.

In this last step we have simply made a rearrangement of the Taylor series so that the indices β_1, β_2 are associated with powers of the real variables x and y. Since $\sigma_i^2(x,y)$ must be real-valued it follows that

$$h_{i\beta_1\beta_2} = 0 \qquad i = 1,2,\ldots,\ell \quad \text{and} \quad \forall\, \beta_1, \beta_2 \geq 0$$

This implies that $\sigma_i^2(x,y)$ has a real Taylor series expansion

$$\sigma_i^2(x,y) = \sum_{\beta_1,\beta_2} e_{i\beta_1\beta_2}\, x^{\beta_1} y^{\beta_2} \tag{A.2.10}$$

which will converge in an appropriate neighbourhood of $(x,y) = (0,0)$. (Recall that we have taken $(s_1^\circ, s_2^\circ) = (s^\circ, \bar{s}^\circ) = (0,0)$.)

Since $\sigma_i^2(x,y)$ is real-analytic by definition and the (positive) square-root of a real-analytic function is a real-analytic function, it follows that $\sigma_i(x,y)$ is real-analytic. To summarize then, the singular-value functions $\sigma_i(x,y)$ of the rational matrix G(s) are real-analytic functions of (x,y), provided $s = x + jy$ is neither a pole nor a "branch point" at which some singular values are equal.

APPENDIX B PROOFS OF PROP 3.2.1, PROP 3.3.1, PROP 4.5.1 AND THEOREM 4.6.2

§B.1 Proof of Prop 3.2.1

We first note that for any $G \in \mathbb{C}^{m\times m}$ (e.g. see [STO])

$$\|G\|_2 \leq \|G\| \leq m^{1/2}\|G\|_2 \tag{B.1.1}$$

Using (3.1.9), (3.1.5) and (B.1.1), we have

$$\begin{aligned} MS(G) &\leq \left(\frac{m^3-m}{12}\right)^{1/4} \Delta(G) \\ &= \left(\frac{m^3-m}{12}\right)^{1/4} \frac{\|GG^*-G^*G\|^{1/2}}{\|G\|} \\ &\leq \left(\frac{m^3(m^2-1)}{12}\right)^{1/4} \frac{\|GG^*-G^*G\|_2^{1/2}}{\|G\|_2} \end{aligned} \tag{B.1.2}$$

Now suppose $Z\Gamma U^*$ is a QND of G (see §3.3) and that $m(G) = \|U^*Z-I\|_2 < \delta$. Then

$$\begin{aligned} \|GG^*-G^*G\|_2 &= \|Z\Gamma^2Z^*-U\Gamma^2U^*\|_2 \\ &= \|U^*Z\Gamma^2-\Gamma^2U^*Z\|_2 \\ &= \|(U^*Z-I)\Gamma^2-\Gamma^2(U^*Z-I)\|_2 \\ &\leq \|U^*Z-I\|_2\|\Gamma^2\|_2 + \|\Gamma^2\|_2\|U^*Z-I\|_2 \\ &= 2\, m(G)\,\|\Gamma\|_2^2 \end{aligned} \tag{B.1.3}$$

Substituting (B.1.3) into (B.1.2) and noting that $\|G\|_2 = \|\Gamma\|_2$, we have

$$\begin{aligned} MS(G) &\leq \left(\frac{m^3(m^2-1)}{3}\right)^{1/4} m(G)^{1/2} \\ &\leq \left(\frac{m^3(m^2-1)}{3}\right)^{1/4} \delta^{1/2} \end{aligned} \tag{B.1.4}$$

Defining $\varepsilon(\delta)$ to be the right hand side of (B.1.4) readily gives the required result (3.2.1) and (3.2.2). □

§B.2 Proof of Prop 3.3.1:

Let $D = \mathrm{diag}(e^{jd_1}, e^{jd_2})$. The squares of the singular values of $(V-D)$ are defined by the equation

$$\det[\nu I - (V-D)^*(V-D)] = 0$$

$$\Leftrightarrow \det[\nu I - (V\Theta^* - D\Theta^*)^*(V\Theta^* - D\Theta^*)] = 0 \tag{B.2.1}$$

Now $$(V\Theta^* - D\Theta^*)^*(V\Theta^* - D\Theta^*)$$

$$= 2I - (V\Theta^*)^*(D\Theta^*) - (D\Theta^*)^*(V\Theta^*)$$

$$= \begin{bmatrix} 2(1-\cos\delta_1\cos\phi) & e^{-j\delta}(e^{-j\delta_1} - e^{j\delta_2})\sin\phi \\ e^{j\delta}(e^{j\delta_1} - e^{-j\delta_2})\sin\phi & 2(1-\cos\delta_2\cos\phi) \end{bmatrix} \tag{B.2.2}$$

where $\delta_1 = d_1 - \theta_1$, $\delta_2 = d_2 - \theta_2$ (B.2.3)

Using (B.2.2) in (B.2.1) and after some simplifications, we have

$$\nu^2 - 4\nu\left(1 - \cos\frac{\delta_1+\delta_2}{2}\cos\frac{\delta_1-\delta_2}{2}\cos\phi\right)$$
$$+ 4\left(\cos\frac{\delta_1+\delta_2}{2} - \cos\frac{\delta_1-\delta_2}{2}\cos\phi\right)^2 = 0$$

The larger of the two roots of this equation is

$$\|V-D\|_2^2 = 2\left[\left(1 - \cos\frac{\delta_1+\delta_2}{2}\cos\frac{\delta_1-\delta_2}{2}\cos\phi\right) + \left|\sin\frac{\delta_1+\delta_2}{2}\right|\sqrt{1 - \cos^2\frac{\delta_1-\delta_2}{2}\cos^2\phi}\right]$$

$$\geq 2(1-\cos\phi) = 4\sin^2\left(\frac{\phi}{2}\right)$$

and equality holds if $\delta_1 = 0 = \delta_2$

$$\Leftrightarrow \quad d_1 = \theta_1, \; d_2 = \theta_2$$

Hence $D = \Theta$ minimizes $\|V-D\|_2$ with the minimum given by

$$\|V-\Theta\|_2 = \left|2\sin\left(\frac{\phi}{2}\right)\right|$$ □

§B.3 Proof of Prop 4.5.1

For notational simplicity, we shall prove the case $m \geq \ell$ only. The following lemma will be needed.

Lemma B.3.1

Suppose $A \varepsilon \mathbb{C}^{m\times\ell}$, $B \varepsilon \mathbb{C}^{\ell\times m}$ $(\ell \leq m)$ are of full rank ℓ.

If

$$BA = \Lambda \qquad (B.3.1)$$

$$AB = \Delta \qquad (B.3.2)$$

where $\Lambda = \mathrm{diag}(\lambda_i)_{i=1}^{\ell}$, $\Delta = \mathrm{diag}(\delta_i)_{i=1}^{m}$ are diagonal marices whose diagonal elements are in descending order of magnitude, then:

(1) Λ is the $\ell\times\ell$ leading submatrix of Δ.

(2) If the diagonal elements of Λ are distinct, then both A, B are pseudo-diagonal (diagonal if $m=\ell$).

(3) If Λ has equal diagonal elements, then both A, B are pseudo-block-diagonal. The sizes of the diagonal blocks are compatible with the number of equal diagonal elements of Λ. Moreover, if $X \varepsilon \mathbb{C}^{t\times t}$ is the diagonal block of A corresponding to t $(\leq \ell)$ equal diagonal elements λ of Λ, then B has a corresponding diagonal block λX^{-1}.

Proof:

(1) Postmultiplying both sides of (B.3.2) by A and using (B.3.1), we have

$$A\Lambda = \Delta A$$

It follows that for any (i,j) s.t. $a_{ij} \neq 0$

$$a_{ij}\lambda_j = \delta_i a_{ij} \qquad (B.3.3)$$

$$\lambda_j = \delta_i \qquad (B.3.4)$$

Since A has full rank, it has at least one nonzero element in each column and so for each λ_j, $\exists\, \delta_i$ such that (B.3.4) holds, which implies that Λ and Δ have the same set of nonzero diagonal elements. If the diagonal elements are also ordered in decreasing order of magnitude, then (1) must be the case.

(2) From (B.3.3), we have

$$(\lambda_j - \delta_i)a_{ij} = 0$$

Now

$$\delta_i = \begin{cases} \lambda_i & i = 1,\dots,\ell \quad \text{by part (1) above} \\ 0 & i = \ell+1,\dots,m \quad \text{since AB has rank } \ell \end{cases}$$

If the λ_i's are distinct, then $\forall i \neq j$

$$(\lambda_j - \delta_i) \neq 0$$

so that we must have $a_{ij} = 0$. Hence A is pseudo-diagonal. A similar argument applies to B.

(3) The proof of (3) is a simple extension of that for (2) and the idea is illustrated by the following simple case.

Let $A \in \mathbb{C}^{4\times 3}$, $B \in \mathbb{C}^{3\times 4}$ and let

$$BA = \Lambda = \mathrm{diag}(\lambda,\lambda,\lambda') \qquad \text{where} \quad \lambda \neq \lambda'$$

$$AB = \Delta = \mathrm{diag}(\lambda,\lambda,\lambda',0)$$

Then by an argument similar to that given in (2), we can show that A, B are of the form

$$A = \left[\begin{array}{cc|c} \multicolumn{2}{c|}{\multirow{2}{*}{X}} & 0 \\ & & 0 \\ \hline 0 & 0 & x \\ 0 & 0 & 0 \end{array}\right] \qquad B = \left[\begin{array}{cc|cc} \multicolumn{2}{c|}{\multirow{2}{*}{Y}} & 0 & 0 \\ & & 0 & 0 \\ \hline 0 & 0 & y & 0 \end{array}\right] \tag{B.3.5}$$

where $X, Y \in \mathbb{C}^{2\times 2}$ are nonsingular and x, y are nonzero. It then follows that $Y = \lambda X^{-1}$. (B.3.6)

□

<u>Proof of Prop 4.5.1</u>:

<u>Necessity</u>:

If both GK and KG are normal, then each is unitarily similar to a diagonal matrix,

$$GK = W\Lambda_{GK}W^* \tag{B.3.7}$$

$$KG = V\Lambda_{KG}V^* \tag{B.3.8}$$

where $W \varepsilon \mathbb{C}^{m\times m}$ and $V \varepsilon \mathbb{C}^{\ell\times\ell}$ are unitary and we assume that the diagonal elements of $\Lambda_{GK}, \Lambda_{KG}$ are put in decreasing order of magnitude. From (B.3.7)

$$W^*GKW = \Lambda_{GK}$$

$$(W^*GV)(V^*KW) = \Lambda_{GK}$$

Similarly, from (B.3.8)

$$(V^*KW)(W^*GV) = \Lambda_{KG}$$

Putting $A = (W^*GV)$ and $B = (V^*KW)$, we can apply Lemma B.3.1 directly. We illustrate the idea behind the rest of the proof by assuming that A, B are of the form (B.3.5).

Now let X have an SVD

$$X = W_1\Sigma_1 V_1^* \tag{B.3.9}$$

then

$$W^*GV = A = \left[\begin{array}{cc|c} \multicolumn{2}{c|}{\multirow{2}{*}{$W_1\Sigma_1V_1^*$}} & 0 \\ & & 0 \\ \hline 0 & 0 & x \\ 0 & 0 & 0 \end{array}\right]$$

so that

$$G = W\left[\begin{array}{cc|c} \multicolumn{2}{c|}{\multirow{2}{*}{W_1}} & 0 \\ & & 0 \\ \hline 0 & 0 & 1 \\ 0 & 0 & 0 \end{array}\right]\left[\begin{array}{cc|c} \multicolumn{2}{c|}{\multirow{2}{*}{Σ_1}} & 0 \\ & & 0 \\ \hline 0 & 0 & x \end{array}\right]\left[\begin{array}{cc|c} \multicolumn{2}{c|}{\multirow{2}{*}{V_1^*}} & 0 \\ & & 0 \\ \hline 0 & 0 & 1 \end{array}\right]V^* \tag{B.3.10}$$

Also from (B.3.5), (B.3.6) and (B.3.9), we have

$$V^*KW = B = \left[\begin{array}{cc|cc} \multicolumn{2}{c|}{\multirow{2}{*}{$V_1(\lambda\Sigma_1^{-1})W_1^*$}} & 0 & 0 \\ & & 0 & 0 \\ \hline 0 & 0 & y & 0 \end{array}\right]$$

so that

$$K = V \left[\begin{array}{cc|c} \multicolumn{2}{c|}{V_1} & 0 \\ & & 0 \\ \hline 0 & 0 & 1 \end{array}\right] \left[\begin{array}{cc|c} \multicolumn{2}{c|}{\lambda\Sigma_1^{-1}} & 0 \\ & & 0 \\ \hline 0 & 0 & y \end{array}\right] \left[\begin{array}{cc|cc} \multicolumn{2}{c|}{W_1^*} & 0 & 0 \\ & & 0 & 0 \\ \hline 0 & 0 & 1 & 0 \end{array}\right] W^* \qquad \text{(B.3.11)}$$

After possibly simple manipulations, we may assume that the right hand sides of (4.5.3) and (B.3.10) are identical decompositions in that

$$Z = W \left[\begin{array}{cc|c} \multicolumn{2}{c|}{W_1} & 0 \\ & & 0 \\ \hline 0 & 0 & 1 \\ 0 & 0 & 0 \end{array}\right] \qquad \Gamma_G = \left[\begin{array}{cc|c} \multicolumn{2}{c|}{\Sigma_1} & 0 \\ & & 0 \\ \hline 0 & 0 & x \end{array}\right] \qquad U^* = \left[\begin{array}{cc|c} \multicolumn{2}{c|}{V_1^*} & 0 \\ & & 0 \\ \hline 0 & 0 & 1 \end{array}\right] V^*$$

Hence (B.3.11) takes the form

$$K = U\Gamma_K Y^*$$

for some diagonal Γ_K, as required.

<u>Sufficiency</u>:

For the given G and K, we have

$$GK = Y\Gamma_G\Gamma_K Y^*$$

and

$$KG = U\Gamma_K\Gamma_G U^*$$

By Theorem 1.6.1, both of these are normal. □

§B.4 Proof of Theorem 4.6.2

(1) By definition

$$\begin{aligned} TPC(Q(s)) &= \lim_{\substack{\delta\to o \\ R\to\infty}} \sum_{i=1}^{m} \Delta_{j\delta}^{jR} \arg \gamma_{Qi}(s) \\ &= \lim_{\substack{\delta\to o \\ R\to\infty}} \Delta_{j\delta}^{jR} \arg\Big(\prod_{i=1}^{m} \gamma_{Qi}(s)\Big) \\ &= \lim_{\substack{\delta\to o \\ R\to\infty}} \Delta_{j\delta}^{jR} \arg(\det Q(s)) \end{aligned}$$

Now as s goes up the imaginary axis (along the Nyquist contour) between 0 and $j\infty$, each pole in $\mathbb{C}_+^*$ ($=\mathbb{C}_+ - \{0\}$) and each zero in $\mathbb{C}_-^\circ$ contributes a net phase change of $\pi/2$ to $\det Q(s)$ while each pole in $\mathbb{C}_-^\circ$ and each zero in $\mathbb{C}_+^*$ contributes a net phase change of $-\pi/2$. Poles and zeros at $s=0$ or $s=\infty$ make no contribution. Hence

$$\mathrm{TPC}(Q(S)) = (P_R - P_L - Z_R + Z_L)\cdot\frac{\pi}{2} \tag{B.4.1}$$

where P_R, P_L and Z_R, Z_L denote the number of poles and zeros of $Q(s)$ in $\mathbb{C}_+^*$, $\mathbb{C}_-^\circ$. Since $Q(s)$ is full rank, we have (e.g. see [KAI, pp.460-461])

$$\begin{aligned}&\text{Total number of (finite + infinite) zeros of } Q(s)\\ =\ &\text{Total number of (finite + infinite) poles of } Q(s)\end{aligned}$$

As $Q(s)$ has no infinite poles because it is proper, we have

$$0 = P_R + P_L + P_o - (Z_R + Z_L + Z_o + \#\mathrm{IZ}(Q(s))) \tag{B.4.2}$$

where P_o, Z_o denote the number of poles and zeros of $Q(s)$ at $s=0$. (B.4.1) and (B.4.2) together gives

$$\mathrm{TPC}(Q(s)) = \left[2(P_R - Z_R) + (P_o - Z_o) - \#\mathrm{IZ}(Q(s))\right]\cdot\frac{\pi}{2}$$

(2) Splitting $(P_R - Z_R)$ into two parts corresponding to $G(s)$ and $K(s)$, we have

$$\begin{aligned}P_R - Z_R = &\left(\#\mathrm{SMP}(G(s),\mathbb{C}_+^*) + \#\mathrm{SMP}(K(s),\mathbb{C}_+^*)\right)\\ &- \left(\#\mathrm{SMZ}(G(s),\mathbb{C}_+^*) + \#\mathrm{SMZ}(K(s),\mathbb{C}_+^*)\right)\end{aligned}$$

Likewise we split $(P_o - Z_o)$ and $\#\mathrm{IZ}(Q(s))$ and after some simple rearrangement, we get

$$\mathrm{TPC}(Q(s)) = \mathrm{TPC}(G(s)) + \mathrm{TPC}(K(s))$$

(3) Under the stated assumptions,

$$\mathrm{TPC}(K(s)) = \#\mathrm{SMP}(K(s),0)\cdot\frac{\pi}{2}$$

which together with (2), gives the required result. □

APPENDIX C THE SYSTEM AUTM

The system AUTM is a 2-input, 12-state, 2-output model of an automobile gas turbine. A state-space description is given in Table C.2. Fig.C.1 shows the Nyquist and Bode magnitude arrays (i.e., the arrays of diagrams consisting of the Nyquist and Bode magnitude plots of the individual elements of the system transfer function matrix).

The open-loop poles and finite/infinite zeros of the system are

Poles	Finite Zeros
-0.216	-0.0956
-0.217	-0.216
-0.218	$-0.233 \pm j1.08$
$-0.370 \pm j1.32$	$-0.923 \pm j1.57$
-0.931	-0.934
-0.933	-8.92
-0.934	-11.1
-8.00	
-9.06	**Infinite Zeros**
-11.0	one 1st order ∞-zero
-11.6	one 2nd order ∞-zero

The system is thus open-loop stable with no zeros in $\mathbb{C}_+$. AUTM has been studied in Examples 2.3.1, 3.2.1, 4.4.1, 5.4.1 and 6.3.1.

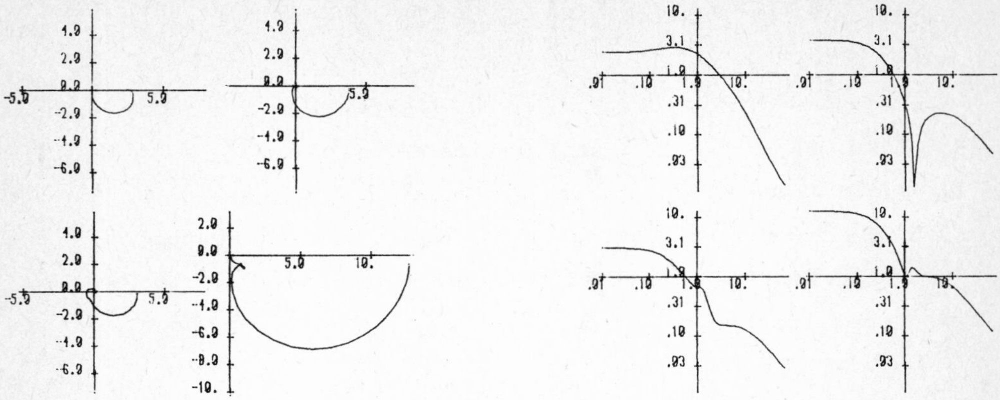

Fig.C.1 Nyquist and Bode magnitude arrays of the system AUTM.

$$A = \begin{bmatrix}
0.0000 & 1.0000 & 0.0000 & 0.0000 & 0.0000 & 0.0000 & 0.0000 & 0.0000 & 0.0000 & 0.0000 & 0.0000 & 0.0000 \\
-0.202 & -1.150 & 0.0000 & 0.0000 & 0.0000 & 0.0000 & 0.0000 & 0.0000 & 0.0000 & 0.0000 & 0.0000 & 0.0000 \\
0.0000 & 0.0000 & 0.0000 & 1.0000 & 0.0000 & 0.0000 & 0.0000 & 0.0000 & 0.0000 & 0.0000 & 0.0000 & 0.0000 \\
0.0000 & 0.0000 & 0.0000 & 0.0000 & 1.0000 & 0.0000 & 0.0000 & 0.0000 & 0.0000 & 0.0000 & 0.0000 & 0.0000 \\
0.0000 & 0.0000 & -2.360 & -13.60 & -12.80 & 0.0000 & 0.0000 & 0.0000 & 0.0000 & 0.0000 & 0.0000 & 0.0000 \\
0.0000 & 0.0000 & 0.0000 & 0.0000 & 0.0000 & 0.0000 & 1.0000 & 0.0000 & 0.0000 & 0.0000 & 0.0000 & 0.0000 \\
0.0000 & 0.0000 & 0.0000 & 0.0000 & 0.0000 & 0.0000 & 0.0000 & 1.0000 & 0.0000 & 0.0000 & 0.0000 & 0.0000 \\
0.0000 & 0.0000 & 0.0000 & 0.0000 & 0.0000 & -1.620 & -9.400 & -9.150 & 0.0000 & 0.0000 & 0.0000 & 0.0000 \\
0.0000 & 0.0000 & 0.0000 & 0.0000 & 0.0000 & 0.0000 & 0.0000 & 0.0000 & 0.0000 & 1.0000 & 0.0000 & 0.0000 \\
0.0000 & 0.0000 & 0.0000 & 0.0000 & 0.0000 & 0.0000 & 0.0000 & 0.0000 & 0.0000 & 0.0000 & 1.0000 & 0.0000 \\
0.0000 & 0.0000 & 0.0000 & 0.0000 & 0.0000 & 0.0000 & 0.0000 & 0.0000 & 0.0000 & 0.0000 & 0.0000 & 1.0000 \\
0.0000 & 0.0000 & 0.0000 & 0.0000 & 0.0000 & 0.0000 & 0.0000 & 0.0000 & -188.0 & -111.6 & -116.4 & -20.80
\end{bmatrix}$$

$$B = \begin{bmatrix}
0.0000 & 1.0439 & 0.0000 & 0.0000 & -1.794 & 0.0000 & 0.0000 & 1.0439 & 0.0000 & 0.0000 & 0.0000 & -1.794 \\
0.0000 & 4.1486 & 0.0000 & 0.0000 & 2.6775 & 0.0000 & 0.0000 & 4.1486 & 0.0000 & 0.0000 & 0.0000 & 2.6775
\end{bmatrix}^T$$

$$C = \begin{bmatrix}
0.2640 & 0.8060 & -1.420 & -15.00 & 0.0000 & 0.0000 & 0.0000 & 0.0000 & 0.0000 & 0.0000 & 0.0000 & 0.0000 \\
0.0000 & 0.0000 & 0.0000 & 0.0000 & 0.0000 & 4.9000 & 2.1200 & 1.9500 & 9.3500 & 25.800 & 7.1400 & 0.0000
\end{bmatrix}$$

$$D = \begin{bmatrix}
0.0000 & 0.0000 \\
0.0000 & 0.0000
\end{bmatrix}$$

APPENDIX D THE SYSTEMS NSRE AND REAC

The system NSRE is a 2-input, 4-state, 3-output "non-square" model of a chemical reactor (see [MUN] and [MAC2]). A state-space description is given in Table D.2. The Nyquist and Bode magnitude arrays are shown in Fig.D.1.

NSRE has open-loop poles at $\{0.0622, 2.01, -5.06, -8.66\}$. The output variables of NSRE have already been arranged in such a way that the first two outputs may be taken as controlled variables. We shall refer to the square system, obtained by simply leaving out the last rows of the C and D matrices of NSRE, as REAC. REAC has finite zeros at $\{-1.19, -5.02\}$ and two 1st order ∞-zeros.

REAC has been considered in Example 3.5.2 and a design for the non-square system NSRE is given in Example 7.3.1.

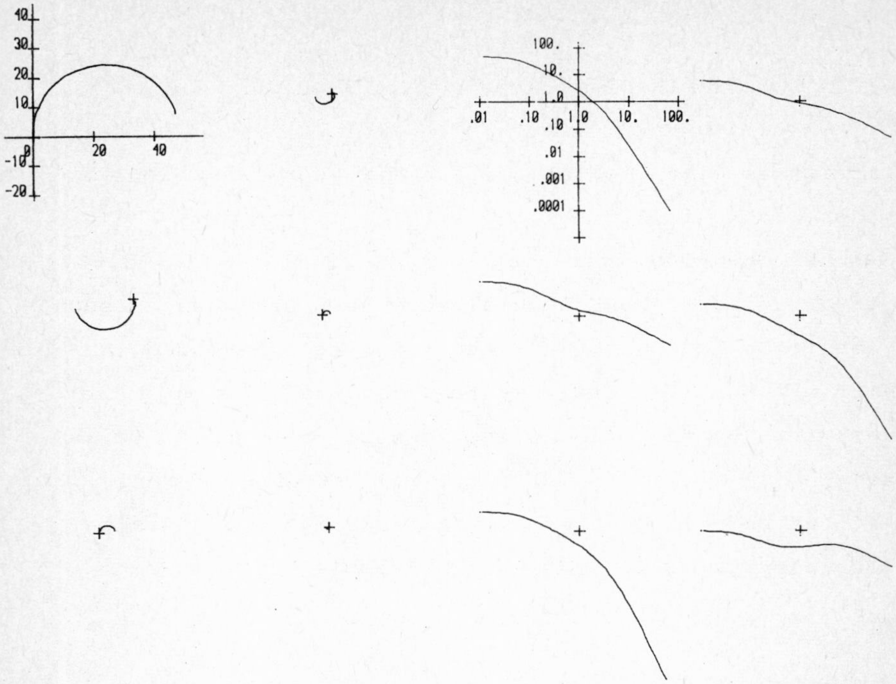

Fig.D.1 Nyquist and Bode magnitude arrays of the system NSRE.
The upper square square block corresponds to REAC.
(All elements are drawn to the same scale as the (1,1)-entry.)

$$A = \begin{bmatrix} 1.4000 & -0.208 & 6.7150 & -5.676 \\ -0.581 & -4.290 & 0.0000 & 0.6750 \\ 1.0670 & 4.2730 & -6.654 & 5.8930 \\ 0.0480 & 4.2730 & 1.3430 & -2.104 \end{bmatrix} \qquad B = \begin{bmatrix} 0.0000 & 0.0000 \\ 5.6790 & 0.0000 \\ 1.1360 & -3.146 \\ 1.1360 & 0.0000 \end{bmatrix}$$

$$C = \begin{bmatrix} 1.0000 & 0.0000 & 1.0000 & -1.000 \\ 0.0000 & 1.0000 & 0.0000 & 0.0000 \\ 0.0000 & 0.0000 & 1.0000 & -1.000 \end{bmatrix} \qquad D = \begin{bmatrix} 0.0000 & 0.0000 \\ 0.0000 & 0.0000 \\ 0.0000 & 0.0000 \end{bmatrix}$$

Table D.2

APPENDIX E THE SYSTEM TGEN

The system TGEN is a 2-input, 10-state, 2-output state-space description of the dynamics (linearized at some nominal operating point) of a turbo-generator (see [LIM]). A listing of the state-space matrices is given in Table E.2 and the Nyquist and Bode magnitude arrays are shown in Fig.E.1.

The open-loop poles and finite/infinite zeros of the system are

Poles	Finite Zeros
-0.231	-1.25
$-0.351 \pm j6.34$	-11.1
-1.04	-20.0
-1.67	$23.0 \pm j425$
-10.0	
-10.7	**Infinite Zeros**
-17.7	one 2nd order ∞-zero
$-29.5 \pm j314$	one 3rd order ∞-zero

A design example for this system is given in §7.1.

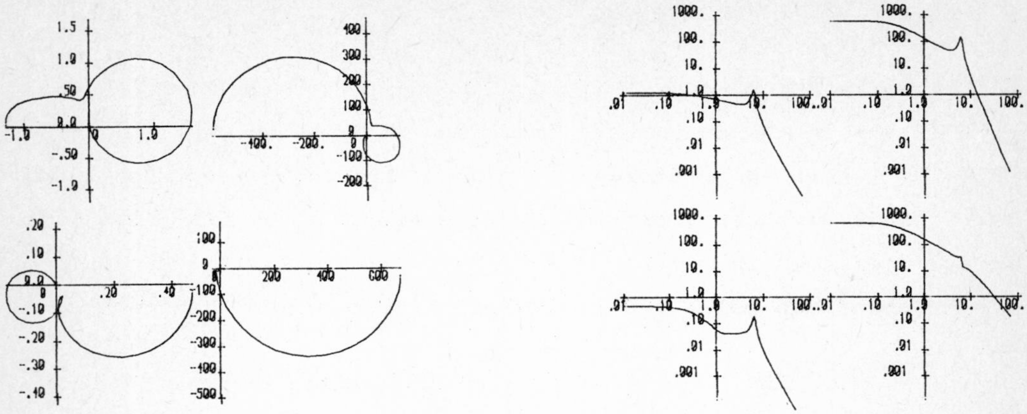

Fig.E.1 Nyquist and Bode magnitude arrays of the system TGEN

$$A = \begin{bmatrix} 0.00000 & 1.00000 & 0.00000 & 0.00000 & 0.00000 & 0.00000 & 0.00000 & 0.00000 & 0.00000 & 0.00000 \\ 0.00000 & -.11323 & -.98109 & -11.847 & -11.847 & -63.080 & -34.339 & -34.339 & -27.645 & 0.00000 \\ 324.121 & -1.1755 & -29.101 & 0.12722 & 2.83448 & -967.73 & -678.14 & -678.14 & 0.00000 & -129.29 \\ -127.30 & 0.46167 & 11.4294 & -1.0379 & 13.1237 & 380.079 & 266.341 & 266.341 & 0.00000 & 1054.85 \\ -186.05 & 0.67475 & 16.7045 & 0.86092 & -17.068 & 555.502 & 389.268 & 389.268 & 0.00000 & -874.92 \\ 341.917 & 1.09173 & 1052.75 & 756.465 & 756.465 & -29.774 & 0.16507 & 3.27626 & 0.00000 & 0.00000 \\ -30.748 & -.09817 & -94.674 & -68.029 & -68.029 & 2.67753 & -2.6558 & 4.88497 & 0.00000 & 0.00000 \\ -302.36 & -.96543 & -930.96 & -668.95 & -668.95 & 26.3292 & 2.42028 & -9.5603 & 0.00000 & 0.00000 \\ 0.00000 & 0.00000 & 0.00000 & 0.00000 & 0.00000 & 0.00000 & 0.00000 & 0.00000 & -1.6667 & 0.00000 \\ 0.00000 & 0.00000 & 0.00000 & 0.00000 & 0.00000 & 0.00000 & 0.00000 & 0.00000 & 0.00000 & -10.000 \end{bmatrix}$$

$$B = \begin{bmatrix} 0.00000 & 0.00000 & 0.00000 & 0.00000 & 0.00000 & 0.00000 & 0.00000 & 0.00000 & 1.66667 & 0.00000 \\ 0.00000 & 0.00000 & 0.00000 & 0.00000 & 0.00000 & 0.00000 & 0.00000 & 0.00000 & 0.00000 & 10.0000 \end{bmatrix}^T$$

$$C = \begin{bmatrix} 1.00000 & 0.00000 & 0.00000 & 0.00000 & 0.00000 & 0.00000 & 0.00000 & 0.00000 & 0.00000 & 0.00000 \\ -.49134 & 0.00000 & -.63203 & 0.00000 & 0.00000 & -.20743 & 0.00000 & 0.00000 & 0.00000 & 0.00000 \end{bmatrix}$$

$$D = \begin{bmatrix} 0.00000 & 0.00000 \\ 0.00000 & 0.00000 \end{bmatrix}$$

Table E.2

APPENDIX F THE SYSTEM AIRC

The system AIRC is a 3-input, 5-state, 5-output model for the vertical-plane dynamics (linearized at datum flight condition) of an aircraft. It is a re-scaled version of the example studied in [KOU] and a listing of the state-space description is given in Table F.2. The Nyquist and Bode magnitude arrays are shown in Fig.F.1. The 5 output measurements are in fact the states of the system, defined as:

x_1 = height error relative to ground or guidance aid, in m;

x_2 = forward speed, in m/sec;

x_3 = pitch angle, in degrees;

x_4 = rate of change of pitch angle, in degree/sec;

x_5 = vertical speed, in m/sec.

The first three states are the variables to be controlled. The inputs are:

u_1 = spoiler angle, in 10^{-1} degrees;

u_2 = forward acceleration due to engine thrust, in m/sec^2;

u_3 = elevator angle, in degrees.

AIRC has open loop poles at $\{0, -0.018 \pm j0.182, -0.78 \pm j1.03\}$, and being non-square, has no zeros. However, if we consider the transmittance from $[u_1\ u_2\ u_3]^T$ to $[x_1\ x_2\ x_3]^T$ (i.e., to the controlled variables), then this square system has zeros given by:

one 1st order ∞-zero

two 2nd order ∞-zeros

no finite zeros

Designs of flight controllers for AIRC are given in Examples 7.3.2 and 7.3.3.

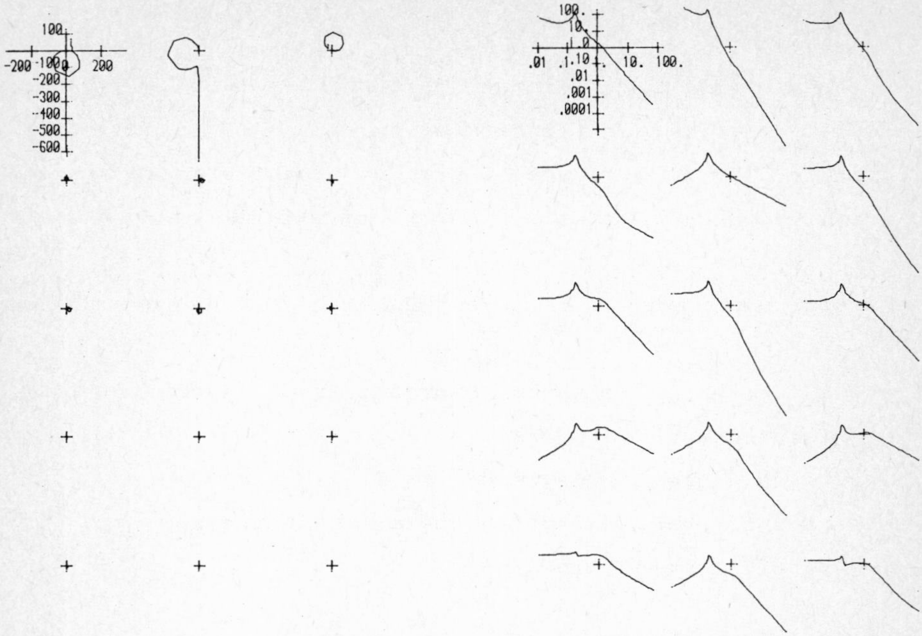

Fig.F.1 Nyquist and Bode magnitude arrays of the system AIRC. (All elements are drawn to the same scale as the (1,1)-entry.)

$$A = \begin{bmatrix} 0.0000 & 0.0000 & 1.1320 & 0.0000 & -1.000 \\ 0.0000 & -.0538 & -.1712 & 0.0000 & 0.0705 \\ 0.0000 & 0.0000 & 0.0000 & 1.0000 & 0.0000 \\ 0.0000 & 0.0485 & 0.0000 & -.8556 & -1.013 \\ 0.0000 & -.2909 & 0.0000 & 1.0532 & -.6859 \end{bmatrix}$$

$$B = \begin{bmatrix} 0.0000 & 0.0000 & 0.0000 \\ -0.120 & 1.0000 & 0.0000 \\ 0.0000 & 0.0000 & 0.0000 \\ 4.4190 & 0.0000 & -1.665 \\ 1.5750 & 0.0000 & -.0732 \end{bmatrix} \qquad C = I_5 \qquad D = O_{5\times 3}$$

Table F.2

REFERENCES

[BEN] A. Ben-Israel & T.N.E. Greville, 1974, Generalized Inverses: Theory and Applications (New York: Wiley).

[BLI] G.A. Bliss, 1966, Algebraic Functions, reprint of 1933 original (New York: Dover).

[BOD] H.W. Bode, 1945, Network Analysis and Feedback Amplifier Design (Princeton: Van Nostrand).

[CRU] J.B. Cruz, J.S. Freudenberg & D.P. Looze, 1981, A relationship between sensitivity and stability of multivariable feedback systems, IEEE Trans. Automat. Contr., vol.AC-26, Feb., pp.66-74.

[CUR] M.L. Curtis, 1979, Matrix Groups (New York: Springer-Verlag).

[DES1] C.A. Desoer & Y.T. Wang, 1980, On the generalized Nyquist stability criterion, IEEE Trans. Automat. Contr., vol.AC-25, pp.187-196.

[DES2] C.A. Desoer & M.J. Chen, 1981, Design of multivariable feedback systems with stable plant, IEEE Trans. Automat.Contr., vol.AC-26, pp.408-415.

[DON] J.J. Dongarra, J.R. Bunch, C.B. Moler & G.W. Stewart, 1979, LINPACK User's Guide (Philadelphia: SIAM).

[DOY] J.C. Doyle & G. Stein, 1981, Multivariable feedback design: Concepts for a classical/modern synthesis, IEEE Trans. Automat. Contr., vol.AC-26, Feb., pp.4-16.

[EDM1] J.M. Edmunds, 1979, Control system design and analysis using closed-loop Nyquist and Bode arrays, Int. J. Control, vol.30, pp.773-802.

[EDM2] J.M. Edmunds, 1981, Cambridge Linear Analysis and Design Package, 3rd Version, Engineering Dept., Cambridge University.

[EDM3] J.M. Edmunds & B. Kouvaritakis, 1979, Extensions of the frame alignment technique and their use in the characteristic locus design method, Int. J. Control, vol.29, pp.787-796.

[FLE] R. Fletcher, 1980, Practical Methods of Optimization, vol.1: Unconstrainted Optimization (New York: Wiley).

[GLA] I.M. Glazman & Ju.I. Ljubic, 1968, Finite-Dimensional Linear Analysis: A Systematic Presentation in Problem Form, English Translation: 1974 (Massachusetts: MIT Press).

[GRA] H. Grauert & K. Fritzsche, 1976, Several Complex Variables, Graduate Texts in Math. (Berlin: Springer-Verlag).

[GAN] F.R. Gantmacher, 1959, Theory of Matrices, vols. I and II (New York: Chelsea).

[GOL] G.H. Golub & V. Pereyra, 1973, The differentiation of pseudo-inverses and nonlinear least-squares problems whose variables separate, SIAM J. Numer. Anal., vol.10, April, pp.413-432.

[HAN] R.J. Hanson & C.L. Lawson, 1974, Solving Least Squares Problems (Englewood Cliffs: Prentice Hall).

[HEN] P. Henrici, 1962, Bounds for iterates, inverses, spectral variation and field of values of non-normal matrices, Numerische Math., vol.4, pp.24-40.

[HUN] Y.S. Hung & A.G.J. MacFarlane, 1981, On the relationships between the unbounded asymptote behaviour of multivariable root loci, impulse response and infinite zeros, Int. J. Control, vol.34, pp.31-69.

[JOH] F. John, 1982, Partial Differential Equations (4th Edition), Applied Mathematical Sciences, vol.1 (New York: Springer-Verlag).

[KAI] T. Kailath, 1980, Linear Systems (Englewood Cliffs: Prentice Hall).

[KAU] L. Kaufman, 1975, A variable projection method for solving separable nonlinear least squares problems, BIT, vol.15, pp.49-57.

[KLE] V.C. Klema & A.J. Laub, 1980, The singular value decomposition: its computation and some applications, IEEE Trans. Automat. Contr., vol.AC-25, April, pp.164-176.

[KOU] B. Kouvaritakis, W. Murray & A.G.J. MacFarlane, 1979, Characteristic frequency-gain design study of an automatic flight control system, Int. J. Control, vol.29, pp.325-358.

[KRO] F. Krogh, 1974, Efficient Implementation of a variable projection algorithm for nonlinear least squares problems, Comm. ACM, vol.17, pp.167-169.

[LAU] A.J. Laub, 1979, An equality and some computations related to the robust stability of linear dynamical systems, IEEE Trans. Automat. Contr., vol.AC-24, April, pp.318-320.

[LEH] N.A. Lehtomaki, N.R. Sandell,Jr. & M. Athans, 1981, Robustness results in linear-quadratic Gaussian based multivariable control designs, IEEE Trans. Automat. Contr., vol.AC-26, Feb., pp.75-92.

[LIM] D.J.N. Limebeer, R.G. Harley & S.M. Schude, 1979, Subsynchronous resonance of the Koeberg turbo-generators and of a laboratory system, Trans. (S.A.) I.E.E., vol.70, pp.278-297.

[MAC1] A.G.J. MacFarlane & I. Postlethwaite, 1977, The generalized Nyquist stability criterion and multivariable root loci, Int. J. Control, vol.25, pp.81-127.

[MAC2] A.G.J. MacFarlane & B. Kouvaritakis, 1977, A design technique for linear multivariable feedback systems, Int. J. Control, vol.25, pp.837-874.

[MAC3] A.G.J. MacFarlane & D.F.A. Scott-Jones, 1979, Vector gain, Int. J. Control, vol.29, pp.65-91.

[MAR] M. Marcus & H. Minc, 1964, <u>A Survey of Matrix Theory and Matrix Inequalities</u> (Boston: Allyn and Bacon).

[MOO] B.C. Moore & L.M. Silverman, 1972, Model matching by state feedback and dynamic compensation, IEEE Trans. Automat. Contr., vol.AC-17, August, pp.491-497.

[MUN] N. Munro, 1972, Design of controllers for unstable multi-variable system using inverse Nyquist array, Proc. IEE, vol.119, No.9, pp.1377-1382.

[MUR] F.D. Murnaghan, 1962, <u>The Unitary and Rotation Groups</u> (Washington: Spartan Books).

[NAR] R. Narasimhan, 1971, <u>Several Complex Variables</u>, Chicago Lectures in Math. (Chicago: University of Chicago Press).

[NUZ] D.W. Nuzman & N.R. Sandell, 1979, An inequality arising in robustness analysis of multivariable systems, IEEE Trans. Automat. Contr., vol.AC-24, June, pp.492-493.

[ORT] J.M. Ortega & W.C. Rheinboldt, 1970, Iterative Solution of Nonlinear Equations in Several Variables (London: Academic Press).

[PEC1] J.L. Peczkowski & M.K. Sain, 1978, Linear multivariable synthesis with transfer functions, in Alternatives for Linear Multivariable Control, edited by M.K. Sain, J.L. Peczkowski & J.L. Melsa, pp.71-87 (Chicago: National Engineering Consortium).

[PEC2] J.L. Peczkowski, M.K. Sain & R.J. Leake, 1981, Multivariable synthesis with inverses, Proceedings Joint Automatic Control Conference, Denver, Colo., pp.375-380 (New York: American Institute of Chemical Engineers).

[POS1] I. Postlethwaite & A.G.J. MacFarlane, 1979, A Complex Variable Approach to the Analysis of Linear Multivariable Feedback Systems, Lecture Notes in Control and Information Sciences 12 (New York: Springer-Verlag).

[POS2] I. Postlethwaite, J.M. Edmunds & A.G.J. MacFarlane, 1981, Principal gains and principal phases in the analysis of linear multivariable feedback systems, IEEE Trans. Automat. Contr., vol.AC-26, Feb., pp.32-46.

[POS3] I. Postlethwaite, 1981, A Design Procedure for linear multivariable feedback systems, Report CUED/F-CAMS/TR-210, Engineering Dept., Cambridge University.

[ROS] H.H. Rosenbrock, 1974, Computer-aided Control System Design (London: Academic Press).

[SAF] M.G. Safonov, A.J. Laub & G.L. Hartmann, 1981, Feedback properties of multivariable systems: The role and use of the return difference matrix, IEEE Trans. Automat. Contr., vol.AC-26, Feb., pp.47-65.

[SAN] N.R. Sandell, 1979, Robust stability of systems with application to singular perturbations, Automatica, vol.AC-15, pp.467-470.

[SMI1] M.C. Smith, 1981, On the generalized Nyquist stability criterion, Int. J. Control, vol.34, November, pp.885-920.

[SMI2] M.C. Smith, 1982, A generalized Nyquist/root-locus theory for multi-loop feedback systems, Ph.D. thesis, Engineering Dept.,

Cambridge University.

[STE1] G.W. Stewart, 1973, Introduction to Matrix Computations (New York: Academic Press).

[STE2] G.W. Stewart, 1973, Error and Perturbation bounds for subspaces associated with certain eigenvalue problems, SIAM Review, vol.15, October, pp.727-764.

[STO] B.J. Stone, 1962, Best possible ratios of certain matrix norms, Numerische Math., vol.4, pp.114-116.

[VER] G.C. Verghese & T. Kailath, 1979, Comments on "On Structural invariants and the root-loci of linear multivariable systems", Int. J. Control, vol.29, pp.1077-1080.

[WAN] S.H. Wang & C.A. Desoer, 1972, The exact model matching of linear multivariable systems, IEEE Trans. Automat. Contr., vol.AC-17, June, pp.347-349.

[WIL] J.H. Wilkinson, 1965, The Algebraic Eigenvalue Problem (Oxford: Clarendon Press).

[WOL] W.A. Wolovich, 1971, The application of state feedback invariants to exact model matching, 5th Annual Princeton Conf. Information Sciences and Systems, Princeton.

BIBLIOGRAPHY

1. Surveys and Collections of Papers

Special Issue on Linear Multivariable Control Systems, 1981, IEEE Trans. Automat. Contr., vol.AC-26, No.1 (Feb.).

Special Issue on Linear Quadratic Gaussian Problem, 1971, IEEE Trans. Automat. Contr., vol.AC-16, No.6 (Dec.).

NATO-AGARD Lecture Series No 117, 1981, Multivariable Analysis and Design Techniques, AGARD, Neuilly-sur Seine, France.

MACFARLANE, A.G.J. (Editor), 1979, Frequency Response Methods in Control Systems, IEEE Reprint Series, IEEE, New York.

MACFARLANE, A.G.J. (Editor), 1980, Complex Variable Methods for Linear Multivariable Feedback Systems, Taylor and Francis, London.

SAIN, M., PECZKOWSKI, J.L. and MELSA, J.I. (Editors), 1978, Alternatives for Linear Multivariable Control, National Engineering Consortium, Chicago.

THALER, G.J. (Editor), 1974, Automatic Control: Classical Linear Theory, Dowden, Hutchinson and Ross, Inc., Stroudsberg, Penn.

2. Classical Feedback

BODE, H.W., 1945, Network Analysis and Feedback Amplifier Design, Von Nostrand, Princeton.

DESOER, C.A., 1965, A general formulation of the Nyquist criterion, IEEE Trans. Automat. Contr., vol.12, pp.230-234.

DESOER, C.A. and VIDYASAGAR, M., 1975, Feedback Systems: Input-Output Properties, Academic Press, New York.

EVANS, W., 1953, Control System Dynamics, McGraw-Hill, New York.

HOROWITZ, I.M., 1963, Synthesis of Feedback Systems, Academic Press, New York.

NYQUIST, H., 1932, Regeneration Theory, Bell Syst. Tech. Journ., 11, pp.126-147.

SMITH, O.J.M., 1958, Feedback Control Systems, McGraw-Hill, New York.

TRUXAL, J.G., 1955, Automatic Feedback Control System Synthesis, McGraw-Hill, New York.

WILLEMS, J.C., 1971, The Analysis of Feedback Systems, M.I.T. Press, Cambridge, Mass.

3. Background Systems and Control Theory

BROCKETT, R.W., 1965, Poles, zeros and feedback: state space interpretation, IEEE Trans. Automat. Contr., 10, pp.129-135.

KAILATH, T., 1980, Linear Systems, Prentice-Hall, Inc., Englewood Cliffs, N.J.

MACFARLANE, A.G.J. and KARCANIAS, N., 1976, Poles and zeros of linear multivariable systems: a survey of the algebraic, geometric and complex-variable theory, Int. J. Control, vol.24, pp.33-74.

MEES, A.I., 1981, Dynamics of Feedback Systems, Wiley, New York.

ROSENBROCK, H.H., 1970, State Space and Multivariable Theory, Nelson, London.

ROSENBROCK, H.H., 1974, Computer-aided Control System Design, Academic Press, New York.

4. Background Matrix and Operator Theory

BEN-ISREAL, A. and GREVILLE, T.N.E., 1974, Generalized Inverses: Theory and Applications, Wiley, New York.

CAMPBELL, S.L. and MEYER, C.D.Jr., 1979, Generalized Inverses of Linear Transformations, Pitman, London.

CURTIS, M.L., 1979, Matrix Groups, Springer-Verlag, New York.

GANTMACHER, F.R., 1959, Theory of Matrices, vol 1 and 2, Chelsea, New York.

GLAZMAN, I.M. and LJUBIC, Ju.I., 1968, Finite-Dimensional Linear Analysis: A Systematic Presentation in Problem Form, M.I.T. Press, Cambridge, Mass.

MARCUS, M. and MINC, H., 1964, A Survey of Matrix Theory and Matrix Inequalities, Allyn and Bacon, Boston, Mass.

STEWART, G.W., 1973, Introduction to Matrix Computations, Academic Press, New York.

STRANG, G., 1976, Linear Algebra and its Applications, Academic Press, New York.

5. Background Complex-Variable Theory

CONWAY, J.B., 1978, Functions of One Complex Variable (2nd edition), Springer-Verlag, New York.

GRAUERT, H. and FRITZSCHE, K., 1976, Several Complex Variables, Springer-Verlag, Berlin.

NARASIMHAN, R., 1971, Several Complex Variables, University of Chicago Press, Chicago.

6. Frequency-Response Methods for Multivariable Systems

EDMUNDS, J.M., 1979, Characteristic gains, characteristic frequencies and stability, Int. J. Control, vol.29, pp.669-706.

EDMUNDS, J.M., 1979, Control systems design and analysis using closed-loop Nyquist and Bode arrays, Int. J. Control, vol.30, pp.773-802.

EDMUNDS, J.M., 1981, Cambridge Linear Analysis and Design Package, 3rd Version, CUED Report, Engineering Dept., University of Cambridge.

EDMUNDS, J.M. and KOUVARITAKIS, B., 1979, Extensions of the frame alignment technique and their use in the characteristic locus design method, Int. J. Control, vol.29, pp.787-796.

HUNG, Y.S. and MACFARLANE, A.G.J., 1981, On the relationships between the unbounded asymptote behaviour of multivariable root loci, impulse response and infinite zeros, Int. J. Control, vol.34, pp.31-69.

MACFARLANE, A.G.J. and KOUVARITAKIS, B., 1977, A design technique for linear multivariable feedback systems, Int. J. Control, vol.25, pp.837-874.

MACFARLANE,A.G.J. and POSTLETHWAITE, I., 1977, The generalized Nyquist stability criterion and multivariable root loci, Int. J. Control, vol.25, pp.81-127.

MACFARLANE, A.G.J. and POSTLETHWAITE, I., 1977, Characteristic frequency and characteristic gain functions, Int. J. Control, vol.26, pp.265-278.

POSTLETHWAITE, I. and MACFARLANE, A.G.J., 1979, A Complex Variable Approach to the Analysis of Linear Multivariable Feedback Systems, Lecture Notes in Control and Information Sciences, vol.12, Springer-Verlag, Berlin.

OWENS, D.H., 1981, Multivariable and Optimal Systems, Academic Press, London.

PATEL, R.V. and MUNRO, N., 1982, Multivariable System Theory and Design, Pergamon Press, Oxford.

SMITH, M.C., 1981, On the generalized Nyquist stability criterion, Int. J. Control, vol.34, pp.885-920.

SMITH, M.C., 1982, A generalized Nyquist/root-locus theory for multi-loop feedback systems, Ph.D. Thesis, University of Cambridge.

7. Performance and Robustness of Multivariable Feedback Systems

DOYLE, J.C. and STEIN, G., 1981, Multivariable feedback design: concepts for a classical/modern synthesis, IEEE Trans. Automat. Contr., vol.AC-26, pp.4-16.

LAUB, A.J., 1979, An equality and some computations related to the robust stability of linear dynamical systems, IEEE Trans. Automat. Contr., vol.AC-24, pp.318-320.

MACFARLANE, A.G.J. and SCOTT-JONES, D.F.A., 1979, Vector gain, Int. J. Control, vol.29, pp.65-91.

MOORE, B.C., 1978, Singular value analysis of linear systems, Part 1 and 2, Systems and Control Reports Nos 7801 and 7802, Electrical Engineering Dept., University of Toronto.

POSTLETHWAITE, I., EDMUNDS, J.M. and MACFARLANE, A.G.J., 1981, Principal gains and principal phases in the analysis of linear multivariable feedback systems, IEEE Trans. Automat. Contr., vol.AC-26, pp.32-46.

SAFANOV, M.G., LAUB, A.J. and HARTMANN, G.L., 1981, Feedback properties of multivariable systems: The role and use of the return difference matrix, IEEE Trans. Automat. Contr., vol.AC-26, pp.47-65.

INDEX

A

Accuracy, tracking, 74
AIRC (aircraft dynamics model), 137-147,169
Algebraic function, 8,150
Aligned, 37
Alignment, 37
  matrix, 16
Analysis, 66
Analytic set, 152
Angular frequency, 5
Approximately, aligned, 40,42
  normal, 19,40,42
AUTM (automobile gas turbine model), 27,29,49,51,99,101,112,114,163

B

Bode, gain-phase relationship, 9
  magnitude array, 163

C

Characteristic gain, 2,150
Characteristic gain loci, 4,7,8
Characteristic values, 4
Characteristic-value decomposition, 2,5
Closed-loop system, 6
Column-Hermite canonical form, 106
Command tracking, 69
Compatibility conditions, 82
Compensator, 4,5
  design, 7
Computer-aided design, 3,66
Conceptual framework, 67
Condition number, 41
Critical frequency range, 72,73
Critical point, 8

D

Design, 7,66
  examples, 117-148
Diagonalizing at a critical frequency, 57
Dimension of matrix group, 26
Discriminant, 152
Disturbance rejection, 69,74

E

Eigenfunctions, 8
Eigenloci, 8
Eigenvalue, 1
  bounds, 18,20,44
Eigenvectors, 1
Encirclements, of the origin, 83
  of the critical point, 8,68
Exact model matching, 90

F

Feedback, configurations, 5,127
  design, 7
Frame matrix, 16
  misalignment, 41
Fréchet derivative, 111
Frequency, 5
Frobenius norm, iv

G

Gain, 1,10
  isotropic, 38
Gain-bandwidth, 9
Gain margin interval, 76,77,87
Gauss-Newton iteration, 110
General linear group, 22,25
Generalized, Nyquist diagram, 1,4
  Nyquist stability criterion, 1,7,8,68
  root-locus method, 1

H

Hermite canonical form, 106

High frequency range, 72,73

I

Indicator, 67
Infinite zero structure, 31,82
Input, frame angles, 27
gain direction, 16
singular-vector frame matrix, 16
space, 5
Interaction, 72
Inverse phase matrix, 16
Inverse return-difference matrix, 70
Irreducible factor, 150

K

Kronecker product, iv,110

L

Least-squares techniques, 90-98,105-112
Left modulus matrix, 16
Left singular vector, 11
Linear least-squares fitting, 90
Linear system, 5
Linear Quadratic Regulator, 143
Low frequency range, 72

M

Matrix groups, 22, 26
Measure, of misalignment, 41,42
of skewness, 19
Minimal-phase, 9
Misalignment, 37
measure of, 41,42
Modulus matrix, 16
Moore-Penrose inverse, 93

N

Negative-unity-feedback configuration, 6
Nonlinear least-squares fitting, 105,129
Non-square systems, 124
Normal, 2,4,17,40
Normality, 16,17,32,37
NSRE (non-square chemical reactor model), 131-136,165
Numerator matrix, 96
calculation of, 91,96
Nyquist, D-contour, 7
generalized diagram, 1,4
generalized stability criterion, 1,7,8
array, 163
-type loci, 27

O

Output, disturbance, 69
frame angles, 27
gain direction, 16
singular-vector frame matrix, 16
space, 5

P

Parametrization, of GL(2,$\mathbb{C}$), 25
of matrix groups, 22,34
of SU(2), 23
of U(2), 24
Parameter group,
decomposition, 21,26,32
loci, 27
Partially gain isotropic, 38
Performance, 1,69
Phase, directions, 17
frame matrix, 17
margin interval, 76,77,87
matrix, 16
Polar decomposition, 2,5,15,17
Postcompensated system, 7
Precompensated system, 6
Principal, gain, 16
gain matrix, 16
phase, 17
Pseudopolynomial, 152

Q

Quasi-classical approach, 3,4
design technique, 66
Quasi-Nyquist, decomposition, 42,43
loci, 3,4,37,49

R

REAC (chemical reactor model), 49,52,165
Real-analytic function, 4,153
Reciprocal M-circles, 78
Regular at infinity, 82
Relative error of fitting, 93
Return-difference matrix, 70
Return-ratio matrix, 70,80
Reversed-frame, 53
design procedures, 91,107
design technique, 4,9
normalization, 80,125
normalizing controller, 4,70,71
Riccati equation, 143
Riemann surface, 4,151
Right modulus matrix, 16
Right singular vector, 11
Robustness, 1,75,80
Roll-off rates, 31,82

S

Schur triangular decomposition, 19,40
Sensor noise rejection, 69
Singular values, 2,3,4,10
analytic properties of, 153
Singular-value, decomposition, 2,10
functions, 3,11,151
Singular vectors, 3,11
Singular-vector-frame misalignment, 41
Skew, 2,19,42
Skewness, measure of, 19,42
Smith-McMillan, poles, v,8,83,84
zeros, v,83,84
Special unitary group, 22,23
Spectral norm, iv,11
Spectral sensitivity, 17,20
Stability, 1,8,68
Standardization, 9,53,54
Subunitary matrix, 11
Synthesis, 66
System description, 5

T

Taylor series, 153,155
TGEN (turbo-generator model), 117-123, 167
Total phase change, 83,84
Total synthesis, 90
Tracking accuracy, 74
Turbo-generator, design example, 117

U

Unitary group, 22,24

W

Winding number, 8